Technical Analysis fi
Recognition

Prodromos E. Tsinaslanidis • Achilleas D. Zapranis

Technical Analysis for Algorithmic Pattern Recognition

Springer

Prodromos E. Tsinaslanidis
The Business School
Canterbury Christ Church University
Canterbury, Kent
United Kingdom

Achilleas D. Zapranis
Department of Accounting and Finance
University of Macedonia
Thessaloniki, Greece

ISBN 978-3-319-23635-3 ISBN 978-3-319-23636-0 (eBook)
DOI 10.1007/978-3-319-23636-0

Library of Congress Control Number: 2015955395

Springer Cham Heidelberg New York Dordrecht London

No information provided in this book should be construed as investment or trading advice or an offer to sell investment advice or any investment product. No representation is being made that any technical trading strategy described within this book will or is likely to achieve profits or losses similar to those shown. Past performance is not indicative of future performance. It should be noted that markets can go up or down and, to our knowledge, there is no perfect technique for investing and trading. So the authors cannot be deemed responsible for any losses arising from the information and tools provided here.

Printed on acid-free paper

Springer International Publishing AG Switzerland is part of Springer Science+Business Media (www.springer.com)

To our families

Preface

Technical analysis is a methodological framework of analyzing, primarily graphically, the historical evolution of financial assets' prices and inferring from this assessment future predictions. Technicians use a variety of technical tools within their trading activities, like filter rules, technical indicators, patterns, and candlesticks. Although most academics regard technical analysis with great skepticism, a significant proportion of practitioners consider technical recommendation within their trading activities. Technical analysis is being used either by academics as an "economic test" of the weak-form efficient market hypothesis or by practitioners as a main or supplementary tool for deriving trading signals.

This book focuses mainly on technical patterns, a topic where existed bibliography usually suffers from critical problems. Books on technical analysis mainly deal with technical indicators, and when referring to patterns, the approach followed is most of times theoretical and descriptive rather than scientific and quantitative. In some cases, only optimal examples are illustrated, which might give the false impression to readers, lacking the required scientific background, that charting is most of the times profitable. Statistical framework for assessing the realized returns is also usually absent. Subjectivity embedded in the identification of technical patterns via visual assessment and various cognitive biases that affect the trading and investment activities of many practitioners place barriers in an unbiased assessment of technical patterns.

The purpose of this book is to deal with the aforementioned problems by approaching technical analysis in a systematic way. This is achieved through developing novel rule-based pattern recognizers and implementing statistical tests for assessing their performance. Our proposed methodology is based on the algorithmic and thus unbiased pattern recognition. The philosophy behind the design of the proposed algorithms is to capture the theoretical principles found in the literature for recognizing visually technical patterns and to quantify them accordingly. The methodological framework we present may prove to be useful for both future academic studies that test the null hypothesis of the weak-form market efficiency and practitioners who want to embed technical patterns within their trading decision-making processes.

Canterbury, United Kingdom Prodromos E. Tsinaslanidis
Thessaloniki, Greece Achilleas D. Zapranis

List of Abbreviations

APT	Arbitrage Pricing Theory
BB	Bollinger Bands
CAPM	Capital Asset Pricing Model
DB	Double Bottoms
DDTW	Derivative Dynamic Time Warping
DT	Double Tops
DTW	Dynamic Time Warping
EMA	Exponential Moving Average
EMH	Efficient Market Hypothesis
GARCH	Generalized Autoregressive Conditional Heteroskedasticity
GBM	Geometric Brownian Motion
HH	Highest High
HS	Head and Shoulders
HSAR	Horizontal Support and Resistance Level
HSARz	Horizontal Support and Resistance Zone
IID	Independent and Identically Distributed
INID	Independent and Not Identically Distributed
IPOCID	Independent Prediction of Change in Direction
LL	Lowest Low
LWMA	Linearly Weighted Moving Average
MA	Moving Average
MAC	Moving Average Crossovers
MACD	Moving Average Convergence Divergence
MAE	Mean Absolute Error
MAPE	Mean Absolute Percentage Error
MOM	Momentum
MSE	Mean Squared Error
NPRMSE	Normalized (by) Persistence Root Mean Squared Error
NRMSE	Normalized Root Mean Squared Error
PIPs	Perceptually Important Points
POCID	Prediction of Change in Direction

POS	Prediction of Sign
PT	Price Target
RB	Rounding Bottoms
RMSE	Root Mean Squared Error
ROC	Rate of Change
RSI	Relative Strength Index
RT	Rounding Tops
RW	Rolling Window
SAR	Support and Resistance
SMA	Simple Moving Average
TA	Technical Analysis
TB	Triple Bottoms
TL	Time Limit
TRB	Trading Range Break-outs
TT	Triple Tops

Contents

Chapter 1
Technical Analysis

1.1 Introduction

Technical analysis (TA) is a methodological framework of analyzing, primarily graphically, the historical evolution of financial assets' prices and inferring from this assessment future predictions. Technicians use a variety of technical tools within their trading activities, like filter rules, technical indicators, patterns and candlesticks. Although most academics regard TA with great skepticism, a significant proportion of practitioners include TA's recommendation within their trading activities.

When typing "technical analysis" in Google (Google scholar) search, results returned are millions (tens of thousands)! Empirical evidences report that 90 % of chief foreign exchange dealers consider technical signals within their investment decisions (Taylor and Allen 1992). Over the years, similar findings for different markets have also been reported which are discussed later in this chapter. The majority of practitioners combine TA with other methodologies, like fundamental and quantitative analysis, for their trading activities with a tendency of using TA for shorter holding periods. Brokerage firms, investment banks and other financial intermediaries also take into consideration TA's investment recommendations in their investment decisions. Today there are numerous software programs and packages dealing with it, whereas journals articles, newsletters and books are myriads. TA is a fact in the making decision process in the financial world and practitioners use it as a main or supplementary tool for deriving trading signals.

Academia has also examined historically, and still does, the efficacy of TA. Particular emphasis has been given on trading systems that include trading rules which can be quantified straightforward like technical indicators. The proportion of studies focusing on technical patterns is minor compared to the massive bibliography which covers TA in general. This can be attributed mainly to the high subjectivity, embedded in the identification and interpretation process of technical

P.E. Tsinaslanidis, A.D. Zapranis, *Technical Analysis for Algorithmic Pattern Recognition*, DOI 10.1007/978-3-319-23636-0_1

patterns. But this subjective nature, the impressive terminology, and probably cognitive biases that affect the trading and investment activities of many practitioners, may all provide reasoning for the apparent impact that TA has in the financial community in general.

Scientific papers primarily use TA as an "economic" test of the weak-form efficient market hypothesis. According to this hypothesis, current prices fully reflect all information embedded in historical prices, making predictions based on historical price evolution a futile attempt. However, although the use of TA is apparent, its usefulness still remains a foggy area, since the majority of academic papers fail to reject the null hypothesis of weak-form efficiency. This puzzling picture is reinforced by the reverse "file drawer" bias according to which "*a researcher who genuinely believes he or she has identified a method for predicting the market has little incentive to publish the method in an academic journal and would presumably be tempted to sell it to an investment bank*"(Timmermann and Granger 2004, p. 15). Based on this excerpt we would like to emphasize that our proposed methodology should not be considered as an investment recommendation by the reader. On the contrary, our empirical evidences suggest that overall TA does not generate systematically, statistically significant abnormal returns.

Our proposed methodology includes the quantification of the descriptive criteria, found in the literature, for technical patterns' visual recognition. This book treats TA systematically, by developing algorithmic, and thus unbiased, technical pattern recognition techniques and implementing a statistical framework for assessing their performance. We believe that our methodology can contribute significantly in future studies that use TA as an economic test of efficient market hypothesis. In addition, the reader will find significant aspects to consider when applying TA in practice. The purpose of this book is presented in more detail later in this chapter.

The rest of the chapter is organized as follows. In Sect. 1.2 the concept of TA is discussed. In Sect. 1.3 the related to TA concepts of Efficient Market hypothesis and the Random Walk hypothesis are presented. Celebrated tools of TA are presented in Sect. 1.4. More precisely, Sect. 1.4.1 discusses on technical indicators, Sect. 1.4.2 focuses on technical patterns, Sect. 1.4.3 refers to candlesticks whilst Sect. 1.4.4 refers to filter rules. Particular emphasis is given to technical patterns since this is the main area this book focuses on. In Sect. 1.5 some controversial perceptions on the characterization and the implementation of TA are presented. More precisely, in Sect. 1.5.1 the science vs. art characterization of TA is discussed, in Sect. 1.5.2 the self-fulfilling prophecy vs. self-destructive nature of TA is presented and finally in Sect. 1.5.3 the back-testing vs. overfitting procedures are discussed. Subsequently, in Sect. 1.6 TA is approached from the perspective of behavioral finance in order to examine its subjective nature. The purpose of the book is described in Sect. 1.7. Finally, in Sect. 1.8, an outline of the book and an overview of each chapter are presented.

1.2 What Is Technical Analysis?

It can be argued that the necessity to observe, interpret, explain and predict is a diachronic characteristic of mankind. On the basis of financial markets, the same feature can be found in the man's effort to explain how financial products' prices evolve and subsequently make predictions. Mainly during the last century, a bundle of technical tools have been developed for this purpose. The cornerstone, on which the whole philosophy of TA is based, is the belief that human beings act similarly under similar conditions. Since investors are humans, this behaviour can be traced in their decision-making processes. TA includes a variety of technical tools which try to classify these repetitive investment/trading behaviours, and their corresponding impacts to the market prices. When price evolves in a known to a technician path, the latter assumes that the price will move with the same manner as it did in the past. Thus, we can initially introduce TA as a process of making predictions of probable future trends, based solely on historical trading experience (open, close, high low prices, volume of transactions etc.). A technician analyses the history of a price series primarily graphically and infers mainly three different trading recommendations: buy (long position), sell (short position) and out of the market (neutral position).

The origins of TA are traced back hundreds of years ago. In the 1600s the candlestick charting was developed by the Japanese in order to analyze the price of rice contracts (Achelis 1995). In west, TA started to gain celebration due to the "Dow Theory" developed by Charles H. Dow who was editor of the Wall Street Journal for the period 1889–1902.[1] Dow developed his theory, but he didn't publish it. Instead, the six tenets of his theory were written in a series of editorials. After his death in 1902, it was his successor at the editorial post William Peter Hamilton who published Dow's work in (1922), and Robert Rhea who expanded it even further in (1932). Today's conventions of TA originate from the principles presented in the Dow Theory.

Below we quote some of the most celebrated definitions found in the literature. These definitions are being cited repetitively in numerous articles. We believe that they can jointly provide a good specification of the concept of TA. The terms "science" and "art" are intentionally highlighted in order to prologue one of the main controversial perceptions on TA which is discussed in more detailed in Sect. 1.5.1.

> Technical Analysis is the study of market action, primarily through the use of charts, for the purpose of forecasting future price trends. The term "market action" includes the three principal sources of information available to the technician—price, volume and open interest. . .chart reading is largely an **art**. (Murphy 1986, pp. 1, 12)

[1] He was the first one to publish a stock market average in 1884. This average was actually the first market index and included 11 stocks. In 1987 the initial index was separated into two indices. The first one included 12 stocks from the industry sector and the second one 20 stocks from the rail sector. For more information concerning Dow's life and the six tenets of the Dow theory see Edwards and Magee (1997).

> Technical Analysis is the **science** of recording, usually in graphic from, the actual history of trading (price, changes, volume of transactions, etc.) in a certain stock or in "the averages" and then deducing from that pictured history the probable future trend. (Edwards and Magee 1997, p. 4)

> The technical approach to investment is essentially a reflection of the idea that prices move in trends that are determined by the changing attitudes of investors toward a variety of economic, monetary, political and psychological forces. The **art** of technical analysis, for it is an **art**, is to identify a trend reversal at a relatively early stage and ride on that trend until the weight of the evidence shows or proves that the trend has reversed. (Pring 2002, p. 2)

Finally, we provide our descriptive definition for the concept of TA:

> We conceive technical analysis as a primarily graphical assessment of the historical evolution of trading-related price paths. Technical tools used for this purpose are designed in the spirit of identifying and classifying, either individually or jointly, the repetitive trading behavior of market participants and its impact on prices. Technicians aim to infer from this assessment information regarding the probable future price progression in terms of support and resistance levels, trend continuations/reversals, turning points and price targets and thus deriving long, short and neutral trading signals.

1.3 Efficient Market Hypothesis

In this section we will briefly present the framework of *Efficient Market Hypothesis* (EMH) along with a variety of related topics. The concept of TA contradicts implications derived from the weak-form of EMH, and thus it can be considered as an economic test for its validity (Campbell et al. 1997).

After the official introduction of the EMH in the late of 70s, this theory gained much attention by the academia and currently, studies, articles, surveys and books referring to it are numerous. According to this hypothesis, a market, for a specific time period, is efficient if prices fully reflect all available information. This implies achieving systematically abnormal returns based on an available set of information is not possible. This statement combined with the innate willingness of mankind to forecast the future resulted in this extensive bundle of articles which test the validity of this hypothesis. Such an attempt can be marked out back in the 5-year empirical research of Cowles (1933) who analysed statistically the forecasting attempts of 45 professional agencies at either picking the most profitable stocks or foretelling the future evolution of the stock market.

During the last decades the concept of EMH is one of the major themes, strongly debated by the academic community. Its origins can be traced in 1900 (Bachelier 1900) in the [as it is referred in Fama (1970)] *ignored for 60 years contribution* of Bachelier who supported that *speculation should be a "fair game"*. As similar statement can be traced in the title of (Samuelson 1965) "*Proof that Properly Anticipated Prices Fluctuate Randomly*".

EMH asserts that current market prices reflect the assimilation of all the information available. Efficiency deals also with the time that new information needs to integrate to market prices. In a market where this hypothesis holds, new information incorporates immediately to current market prices and adjusts them to "fair" levels. The decade of 70s is considered the starting point in the modern economic literature of the EMH with the article of Fama (1970) who states that *a market in which prices always "fully reflect" available information is called "efficient"*. The mathematical groundwork though, had been forth earlier presented in Mandelbrot (1966). EMH asserts that market prices fully reflect all the information available, and thus it is unfeasible for someone to realize systematically, statistically significant abnormal returns. Fama expresses this stochastic behavior of market prices as:

$$E\left(\tilde{P}_{j,t+1}|\Phi_t\right) = \left[1 + E\left(\tilde{r}_{j,t+1}|\Phi_t\right)\right]P_{j,t}, \tag{1.1}$$

where $P_{j,t}$ is the price of security j at time t; $r_{j,t+1}$ is the one-period arithmetic return i.e., $\left(P_{j,t+1} - P_{j,t}\right)/P_{j,t}$; Φ_t is the information set that is assumed to be "fully reflected" in the price at time t; and the tildes indicate a random variable. On the basis of (1.1), Fama states that this rules out the possibility of trading systems based solely on information set Φ_t that have expected profits or returns in excess of equilibrium expected profits or returns. Quantitatively this is:

$$E\left(\tilde{z}_{j,t+1}|\Phi_t\right) = 0, \tag{1.2}$$

where $z_{j,t+1}$ is the difference between the observed price (or return) and the expected value of the price (or return) of security j, that was projected at time t on the basis of the information set Φ_t. In other words the sequence $\{z_{j,t}\}$ is a "fair game"[2] pertaining to the information sequence $\{\Phi_t\}$. Due to (Roberts 1967), Fama identifies three different forms of efficiency (weak, semi-strong and strong form) regarding three different subsets of available information sets: historical prices, public available and private information respectively. In addition to historical prices, variables like dividend yields and interest rates were added to this subset of information in Fama (1991).

The existence of weak form efficiency implies that no investor can achieve systematically abnormal returns by using information embedded in historical prices. The term "systematically" is necessary in the previous statement so as to exclude the cases where an investor can "beat" the market by chance. In a weakly efficient market, TA and every trading strategy based solely on historical prices is not expected to embed any systematic predictive power.

Publicly available information subset is used to decide if a financial market can be characterized as semi-strong form efficient or not. This form of efficiency includes the weak-form, which means that when a market is semi-strong efficient it must be also weak-form efficient, while the reverse does not necessarily hold. In addition, if a market is semi-strong efficient the potential implications of publicly available information are already reflected in market prices. Thus, neither technical

[2] See martingale model discussed later in this section.

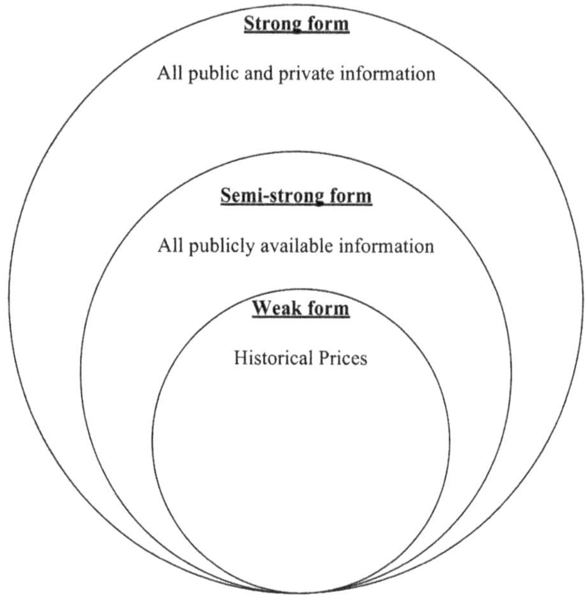

Fig. 1.1 Diagram presenting the relationship between the three forms of EMH. The corresponding subsets of information used for defining each form are also presented

analysis nor fundamental analysis is worth to trust for investment and trading decisions.

Private information refers to the inside (not publicly available) information and the special information that investors can extract by their own analysis. In a strong form efficient market no one can realize systematically excess returns by taking advantage of any subset of information, private or publicly available. In Fig. 1.1 it is obvious that this case of efficiency is the wider of all and includes both the aforementioned forms of efficiency.

Random walk hypothesis and martingale models are closely related with the EMH. In gambling theory a martingale model characterizes a "fair game" where chances of winning are neither in our favour nor ours opponents. The "fair game" model for a weak form efficient market implies no predictability for futures returns (or prices), and constitutes the basis on which the random walk models are being erected.

As we already mentioned, in a weak form efficient market, no trading system based on historical observations can predict future price changes in an accurate manner. In probability theory, this statement is enclosed in the martingale model. Quantitatively a martingale model is a stochastic process which fulfills the following conditions:

$$E\left[P_{t+1}\middle| P_t, P_{t-1}, P_{t-2}, \cdots\right] = P_t \tag{1.3}$$

or equivalently,

$$E\left[P_{t+1} - P_t\middle| P_t, P_{t-1}, P_{t-2}, \cdots\right] = 0 \tag{1.4}$$

A martingale is thus a sequence of random variables $\{P_t\}$ for which at a specific time t in the realised sequence, the next value's expectation $E[P_{t+1}]$ equals with the current observed value, conditioned on all preceding observed values (1.3) Equivalently, the expectation for a subsequent future price change equals with zero (1.4). In view of finance applications, if P_t is the price of a financial asset at time t, (1.3) and (1.4) imply that the best expectation of tomorrow's price (or price change) is today's price (or zero) given the historical price path.

However, the martingale model does not consider risk-premiums embedded in price returns. Campbell et al. (1997, p. 31) quote: "… *if an asset's expected price change is positive, it may be the reward necessary to attract to hold the asset and bear its associated risks. Therefore … it has been shown that the martingale property is neither necessary nor a sufficient condition for rationally determined asset prices*". Even so, the martingale model had been the cornerstone on which the random walk hypothesis was build.

Three are the most celebrated versions of the random walk hypothesis in the bibliography dubbed *RW*1, *RW*2 and *RW*1. Random walk models describe the manner that financial price series evolve and these three models differ in the assumptions underlying the corresponding model's increments. As we proceed from the first to the last model these assumptions are being relaxed gradually, providing each model a wider nature compared to each previous.

When adjusted to financial assets' price series, the first version (*RW*1) is given by (1.5).

$$p_t = \mu + p_{t-1} + e_t, \qquad e_t \sim IID(0, \sigma^2) \tag{1.5}$$

Here the natural logarithm of prices $(p_t \equiv \ln P_t)$ is defined by the sum of a constant μ, the natural logarithm of the prior observed price p_{t-1} and a random shock e_t.[3] Constant expresses the expected return of the random variable (or drift rate) and increments are *independent and identically distributed* (IID) with a mean of zero and variance σ^2.

A wider version of *RW*1 is the *RW*2, which results by relaxing the prior assumption that residuals are identically distributed. By this manner, the model represents the evolution of real price series in a more efficient way since it allows for unconditional heteroskedasticity. Financial asset's return series embed this phenomenon, while they exhibit time-varying volatility clustering. Thus, under the *RW*2 framework e_t are *independent and not identically distributed* (INID) and (1.5) becomes:

$$p_t = \mu + p_{t-1} + e_t, \qquad e_t \sim INID(0, \sigma_t^2) \tag{1.6}$$

[3] We use the natural logarithm of prices instead of the actual prices, because in the latter case, when assuming normal distribution for the variable P_t, there is always a positive probability to observe a negative price, which is not possible.

Finally an asset's return series is said to follow a *RW*3 when increments are dependent (i.e., $cov(e_t^2, e_{t-k}^2) \neq 0$ for some $k \neq 0$) but uncorrelated [$cov(e_t, e_{t-k}) = 0$ for all $k \neq 0$].[4] Several methods have been proposed in the bibliography to test the random walk hypothesis [see Campbell et al. (1997) and references therein]. This book focuses on TA, which is considered a powerful tool to perform an economic test of *RW*2 hypothesis and thus the weak form EMH.

1.4 Celebrated Tools of Technical Analysis

Technical oriented investors have designed an assortment of technical tools to monitor historical price paths and infer, from this mainly graphical assessment, predictions. In this section we present four celebrated classes of technical tools, namely technical indicators, patters, candlesticks and filter rules. Particular emphasis is given on technical patterns which is one of the main areas this book covers.

1.4.1 Technical Indicators

Technical indicators are the outputs derived by a set of mathematical formulas which use trading-related historical data as inputs. For their calculation, the user is required to define a number of parameters, like the requested time period. Performing TA by using this type of tools involves drawing them either on or above the chart. Examples of well-known technical indicators are *moving averages* (MA), *moving average convergence divergence* (MACD), *relative strength index* (RSI), *Bollinger bands* (BB), *momentum* (MOM) etc which are covered in more detailed in Chap. 7. Recall from the descriptive definition of TA we provided in Sect. 1.2, we emphasized that technical tools are used "individually or jointly". This is aligned with the excerpt from Pring (2002, p. 11) who states "*No single indicator can ever be expected to signal all trend reversals, so it is essential to use a number of them together to build up a consensus*". In addition, the existence of any potential evidence in favour of an individual technical indicator or pattern should not be regarded as a validation of TA's predictability. There is a wide range of technical indicators and patterns and the generalised value of TA should be estimated on the basis of this universe. Otherwise the predictive power of one technical rule, at one market for a specific time period could be attributed to pure chance. In Chap. 8 we assess the predictive performance of the technical patterns whose identification

[4] *cov* stands for covariance. For a more detailed and comprehensive description of the three versions of random walk hypothesis along with their corresponding tests, see Campbell et al. (1997).

process is presented in Chaps. 4–6 along with a bundle of technical tools that will be presented in Chap. 7.

1.4.2 Technical Patterns

Technical patterns are specific forms of evolutions of price paths which are mainly identified visually. When such a pattern is identified, a technician expects prices to evolve with a specific way. Although descriptions for defining such patterns can be found in numerous technical manuals, web-pages and so on, this pattern recognition process embeds a high level of subjectivity. This subjectivity and various cognitive biases, like clustering illusion (see Sect. 1.6) reported in the context of behavioral finance, place frontiers in the assessment of the efficacy of such patterns. Scientific articles testing the predictive power of chart patterns constitute only a minor fraction of the existing bibliography on testing the profitability of TA in general. This dismal size of the bibliography can be mainly attributed to the difficulty that lies in the development of efficient and robust pattern identification mechanisms that integrate all relevant subjective principles of TA, in an objective manner. The scope of this book is to shed light in this foggy area of TA. To do so, we propose novel rule-based identification mechanisms for technical patterns. We generally classify them into three different categories: horizontal, zig-zag and circular patterns. The proposed algorithmic identification mechanisms are presented in Chaps. 4–6. This section presents representative studies with their empirical results for the technical patterns considered in this book.

A number of surveys examined the efficacy of assembles of technical patterns. Preliminary evidence for the lack of predictive power of 32 five-points chart patterns were presented in Levy (1971). The dataset used was daily, adjusted, closing prices of 548 securities from the NYSE. At the period July 3 1964 through July 4 1969, none of the 32 five-points chart patterns produced better than average profits for the afore mentioned securities. In Lo et al. (2000) a pattern identification mechanism based on kernel regression was developed. The later was used to smooth price series and identify regional peaks and bottoms. With the fulfillment of specific criteria the identification mechanism recognised various technical patterns including the *head and shoulders* (HS) pattern. When applied on US stocks for the period 1962–1996 they found that there is potentially useful information embedded in most of the technical patterns they consider, although this does not necessarily mean that TA can generate trading profits. Dawson and Steeley (2003) employed Lo et al.'s method on UK stocks over the period 1989–2001. Their results are marginally weaker than the ones found in Lo et al. (2000), but they are generally aligned with them. More concretely, the distributions of returns conditioned on the examined technical patterns were significantly different from the unconditional ones, although means of those distributions were not. Hsu and Kuan (2005) examined a relative wide "universe" of technical trading rules based on either isolated simple rules or complex trading strategies. In their test they considered data snooping bias by implementing White's reality check (White 2000) and Hansen's SPA test (Hansen 2005). Within their results they found that TA is profitable in relatively "young" markets (NASDAQ Composite and Russell 2000) but not in relatively "mature" markets (DJIA and S&P 500). The comparison of trading results was made with buy-and-hold strategies taking into consideration transaction costs (0.05 % for each one-way trade) in a manner similar with the one used in Fama and Blume (1966).

Carol Osler, while working in Federal Reserve Bank of New York, investigated the predictive performance of the HS pattern and *support and resistance* (SAR) levels. Concerning the later technical tool, Osler (2000) used published SAR levels provided by six firms, for three currency pairs (dollar-mark, dollar-yen and dollar-pound) for the period 1996–1998. She examined the profitability of those levels on intraday exchange rates, using a bootstrapped technique. Among her results, she mentioned that the published levels outperform the ones created artificially, a finding in favour of the efficacy of TA. In contrast, strengths of the published levels, provided by three of those firms were found to be insignificant. In addition, in (2002), she exhibited that exchange rates do tend to reverse trend at round numbers, a price behaviour accepted by technicians. In an effort to explain this behaviour, in Osler (2003) it was found that take-off orders seem to cluster strongly at round numbers whereas stop-loss orders appear to cluster just beyond round numbers. The first finding corroborates the belief of technical community that trends tend to reverse course at predictable SAR levels, while the second one the confirms the expectation that trends move unusually rapidly after prices penetrate such levels. The connection between stop-loss orders and rapid, self-reinforcing price movements was examined in Osler (2005). Curcio et al. (1997) examined the profitability of SAR levels appeared on the FXNL screen on intra-daily foreign exchange markets. Their main result was that even when transaction costs were not taken into account, it was on average not profitable to trade according to these rules, although excess returns could be made in periods of strongly trending exchange rates. Finally empirical results regarding *trading range break-outs* (TRB) are numerous, probably due to the simplicity in their identification process.[5] Due to this fact, this variation of SAR levels is usually assessed with a bundle of other, easily identified technical indicators and results are mixed (Brock et al. 1992; Cheung et al. 2011; Coutts and Cheung 2000; De Zwart et al. 2009; Falbo and Cristian 2011; Hudson et al. 1996; Marshall et al. 2008a, b; Marshall et al. 2009; Mizrach and Weerts 2009; Qi and Wu 2006; Raj and Thurston 1996).

Regarding the HS pattern, one of the first attempts to quantify this technical pattern with a rule-based system was presented in Osler and Chang (1995). In this study, several precise criteria were set, whose fulfillment was a necessary condition for the pattern's confirmation. Clyde and Osler (1997) use the HS pattern to provide a theoretical foundation and supportive evidence for the linkage (or equivalence)

[5] This type of SAR levels is identified from minima or maxima realized in a precedent, constant time intervals (see Chap. 4).

between TA and nonlinear forecasting on high dimension systems. Osler (1998) states that investment positions based on signals generated by visual price patterns constitute a source of noise trading. The reason for this impact is the fact that trading volume realized during the pattern's formation is unusually high, and trading based on the signals generated are unprofitable. More concretely, she focused on the HS pattern and she found that the pattern was not profitable when applied on 100 randomly selected firms for the period 1962–1993. Chang and Osler (1999) used the HS pattern on daily dollar exchange rates over the period 1973–1994. The pattern was found to be profitable for the mark and yen, but not for the other four exchange rates (Canadian dollar, Swiss franc, French franc and U.K. pound). In those profitable cases the comparison was made with both the average buy-and-hold mark's return and the corresponding average annual return of S&P 500. This profitability is also robust even when the comparison is made with two null models; Random walk and GARCH (1, 1) processes. However, they state "*the HS trading rule is profitable but not efficient as it is dominated by other simpler trading rules (moving averages)*". These simpler trading rules are much easier for an investor to calculate and thus there is no reason to include HS's signals into his investment decision process, even in the mark and yen currencies. In their last comment they assert that the persistence on using the HS pattern constitutes a cause of predictable exchange-rate forecast errors. In addition, Neftci (1991) devised some formal algorithms to represent various forms of TA and he showed that a few of those rules generate well-defined techniques of forecasting. Lucke (2003) used the criteria presented in Osler and Chang (1995) for the identification of the HS pattern. The pattern was tested on daily spot exchange rates for five currencies, covering all relevant cross rates. Returns generated were not significantly positive. In a more recent study Savin et al. (2007) used a modified version of the pattern recognition algorithm presented in Lo et al. (2000) in order to evaluate the predictive power of the HS pattern. Their modifications concern a different span of the rolling window implemented, a different bandwidth and some additional restrictions on the locals' criteria. The last modification is based on technical principles found in Bulkowski (2000). The efficacy of the HS pattern was examined in the S&P 500 and the Russell 2000 components over the period 1990–1999 and they found no profitability on a trading strategy based solely on the examined pattern. In contrast they found that HS patterns could be used successfully in conjunction with a passive indexing strategy to improve the trade-off between risk and return. Friesen et al. (2009) state that confirmation bias embedded in investors' "decision-making" procedures generates autocorrelation and patterns of price evolution that can predict future prices. Their analysis provides reasoning on the existence of the HS and double-tops patterns. In Zapranis and Tsinaslanidis (2010a) the criteria mentioned in Lucke (2003) were applied through a rule based mechanism on stochastically generated price series in an attempt to view through the prism of behavioral finance the HS pattern's performance. The HS pattern is successfully identified in random price series and this indicates that it is possible the pattern to be identified in real price series too. The main conclusion is that if the geometric Brownian motion is considered an accurate representation of the price stock generating mechanism then the HS pattern has no predictive power at all. An alternative way of identifying the HS pattern with artificial neural networks was also proposed in Zapranis and Tsinaslanidis (2010b).

Empirical results on the *rounding bottoms* (RB) pattern (also known as “saucers” in technical jargon) are very limited due to specificities that characterize this pattern. In contrast with most zigzag technical patterns, this one is circular, and more complicated techniques for its identification are necessary to be developed. Some studies propose identification mechanisms for this pattern based on criteria applied on regional peaks and bottoms (Hsu and Kuan 2005; Lo et al. 2000). An alternative identification approach was followed by Wang and Chan (2009) who use a template matching technique (Duda and Hart 1973) to identify the rounding top and saucer pattern. They state that the pattern has considerably forecasting power across various tech stocks traded in U.S. Chapter 6 presents our proposed identification mechanism, which captures the circular characteristics of this pattern (Zapranis and Tsinaslanidis 2012b).

In Leigh et al. (2002a, b, c) template matching techniques, for the identification of “bull-flag” variations, are being used. They examined the profitability of these trading rules on NYSE composite index, and they found that these rules are efficient. The same patterns were used in order to identify abrupt increases in volume in NYSE (Leigh et al. 2004). Such volume increases were found to be followed by price increases under certain condition during the examined period (1981–1999). An alternative template matching technique was implemented in Bo et al. (2005), which they claim outperforms the classic one. Their results concern the Shanghai Stock Exchange and their evidence are against the null hypothesis of EMH. The aforementioned studies were extended for the NASDAQ Composite Index and Taiwan Weighted Index in Wang and Chan (2007). Finally, Leigh et al. (2008) extended their previous works and reported results supporting the predictive power for the “bear-flag”, which is the mirrored pattern of the “bull-flag”.

At this point the aforementioned studies are classified into four different categories according to the patterns they examined and some of the most representative ones are presented in the following tables. More precisely, selective bibliography on assessing sets of mixed technical trading rules including technical patterns is presented in Table 1.1. Table 1.2 presents selective bibliography on technical trading rules based on horizontal patterns. Table 1.3 presents selective bibliography on technical trading rules based on zigzag patterns. Finally, Table 1.4 presents selective bibliography on technical trading rules based on circular patterns.

1.4.3 Candlesticks

A candlestick illustrates the price evolution of a financial asset for a particular time span. Its drawing requires the open, close, high and low price observed during the requested period. Figure 1.2 illustrates the characteristic components of a

Table 1.1 Selective bibliography assessing sets of mixed technical trading rules including technical patterns

Study	Pattern(s) examined	Dataset	Main empirical findings/ contributions/ conclusions
Levy (1971)	32 five-point chart formations	548 NYSE stocks for the period 1964–1969	1) "*. . .after taking trading costs into account, none of the thirty-two patterns showed any evidence of profitable forecasting ability in either (bullish or bearish) direction*". (p. 318) 2) "*Moreover, the most bullish results tended to be generated by those patterns which are classified as bearish in the standard textbooks on charting, and vice versa*". (p. 318)
Lo et al. (2000)	10 chart patterns: HS, inverse HS, broadening tops and bottoms, triangle tops and bottoms, rectangle tops and bottoms, and double tops and bottoms	US stocks for the period 1962–1996	"*Several technical indicators do provide incremental information and may have some practical value*". (p. 1705)
Dawson and Steeley (2003)	Similar with (Lo et al. 2000)	UK stocks for the period 1989–2001	Similar with (Lo et al. 2000) but marginally weaker results
Hsu and Kuan (2005)	39,832 rules and strategies	Four market indices: DJIA, S&P 500, NASDAQ composite and Russell 2000 for the period 1989–2002	"*. . .significantly profitable simple rules and complex trading strategies do exist in the data from relatively "young" markets (NASDAQ Composite and Russell 2000) but not in the data from relatively "mature" markets (DJIA) and S&P 500)*". (p. 606)

Italics signify excerpts from the corresponding studies

candlestick. In technical jargon the main part of the candlestick is called the real body. It can be hollow (left) or filled (right) depending whether the open price is lower or greater than the close. These prices are signified by the upper and lower edge of the real body accordingly. The thin lines above and below the real body are called upper and lower shadows respectively. The top (bottom) of the upper (lower)

Table 1.2 Selective bibliography assessing technical trading rules based on horizontal patterns

Study	Pattern(s) examined	Dataset	Main empirical findings/ contributions/conclusions
Curcio et al. (1997)	1) Support and resistance levels 2) high-low trading ranges 3) Combinations of 1 and 2 4) Max-min ranges from Brock et al. (1992) 1 and 2 were downloaded from the Reuters FXNL screen.	Intra-daily foreign exchange markets (Mark, Yen and British Pound against US Dollar) for two sample periods (10 April 1989–29 June 1989 and 31 January 1994–30 June 1994)	"*... even when transaction costs were not taken into account, it was on average not profitable to trade according to these rules, although excess returns can be made in periods of strongly trending exchange rates*" (p. 269)
Osler (2000)	Published support and resistance levels provided by six firms	Intraday data for three currency pairs (dollar-mark, dollar-yen and dollar-pound), for the period Jan 1996–March 1998	1) Published SAR levels outperform the ones created artificially. 2) "*In short, published estimates of the strength of the levels do not seem to be useful*" (p. 64)
Osler (2002)	Support and resistance levels	Intraday data for three currency pairs (dollar-mark, dollar-yen and dollar-UK pound), for the period Jan 1996–April 1998	Exchange rates change direction when they reach round numbers more frequently than they do at other numbers
Osler (2003)	Support and resistance levels	Stop-loss and take-profit orders for three currency pairs: dollar-yen, dollar-UK pound and euro-dollar for the period 1 August 1999–11 April 2000	"*Take-profit orders cluster particularly strongly at round numbers*"... whereas ..."Stop-loss orders cluster strongly just beyond round numbers" (p. 1791)
Zapranis and Tsinaslanidis (2012a)	Horizontal support and resistance levels	232 stocks from NASDAQ and 501 stocks from NYSE (1990–2010)	"*...support levels outperform resistance ones in predicting trend interruptions but they fail to generate excess returns when they are compared with simple buy-and-hold strategies*". (p. 1571)

Italics signify excerpts from the corresponding studies

shadow mark the high (low) price observed during the requested by the user time span.

A candlestick pattern is identified when an individual or a sequence of candlesticks with specific characteristics is observed. There are various candlestick

Table 1.3 Selective bibliography assessing technical trading rules based on zig-zag patterns

Study	Pattern(s) examined	Dataset	Main empirical findings/ contributions/conclusions
Neftci (1991)	HS and triangles	Markov Times and Dow-Jones industrials for the period 1911–1976	He compared technical analysis methods to Wiener–Kolmogorov prediction theory. He concluded: "*If the processes under consideration were nonlinear, then the rules of technical analysis might capture some information ignored by Wiener-Kolmogorov prediction theory*" (p. 570) In addition he devised some formal algorithms to represent the examined technical patterns
Osler and Chang (1995)	HS	Daily exchange rates of major currencies versus the dollar, for the period March 1973—June 1994	They proposed a set of criteria for the identification of the HS pattern
Clyde and Osler (1997)	HS	Chaotic series	They provided a theoretical foundation and supportive evidence for the linkage (or equivalence) between technical analysis and nonlinear forecasting on high dimension systems
Osler (1998)	HS	Daily trading volume and prices for 100 US firms chosen Randomly from the CRSP database for the period 1962–1993	No profitability for the HS pattern
Chang and Osler (1999)	HS + indicators (oscillators and momentum rules)	Daily spot rates for six currencies versus the US dollar for the period 19 March 1973–13 June 1994	"*The rule is profitable but not efficient, since it is dominated by simpler trading rules*" (p. 363)
Leigh et al. (2002a, b, c)	Bull flag variations	NYSE composite index	Indications against EMH
Lucke (2003)	HS	Daily spot exchange rates	1) "*Returns to SHS trading are not significantly positive—and if there is any evidence for non-zero returns at all, then it is evidence for negative returns*" (p. 39) 2) "*...the strongest evidence for excess profitability is found for the shortest holding period*" (p. 37)

(continued)

Table 1.3 (continued)

Study	Pattern(s) examined	Dataset	Main empirical findings/ contributions/conclusions
Leigh et al. (2004)	Bull flag for trading volume	NYSE composite Index for the period 1981–1999	"*Such volume increases are found to signal subsequent increases in price under certain conditions during the period from 1981 to 1999, the Great Bull Market*" (p. 515)
Bo et al. (2005)	Bull Flag	Shanghai Stock Exchange Composite Index for the period 1993–2004	Indications against EMH
Savin et al. (2007)	HS	S&P 500 and Russell 2000 for the period 1990–1999	The provided evidence in favour of the HS pattern's predictive power
Leigh et al. (2008)	Bull/Bear Flag	NYSE Composite Index for the period (1967–2003)	Indications against EMH
Friesen et al. (2009)	HS and DT	Quotes from NYSE NASDAQ and AMEX for the period 1999–2005	They state that confirmation bias produces autocorrelations and thus it reasons for the existence of technical patterns like HS and DT

Table 1.4 Selective bibliography assessing technical trading rules based on circular patterns

Study	Pattern(s) examined	Dataset	Main empirical findings/ contributions/conclusions
Wang and Chan (2009)	Rounding Tops/ Bottoms	Seven US tech stocks for different sets of holding periods	1) Identification of the pattern with a template matching technique 2) Indications against EMH
Zapranis and Tsinaslanidis (2012b)	Rounding Tops/ Bottoms and resistance levels	Same dataset with Wang and Chan (2009)	1) Identification of the rounding bottoms and resistance levels with a rule based mechanism 2) Resistance levels outperform rounding bottoms Statistical significant excess returns are being generated mainly in the first sub-period considered. They subsequently decline or even vanish as more recent years are examined

patterns acknowledged in TA community with impressive names, like "*the hammer*", "*the shooting star*" and "*doji*". However, this tool of TA is out of the scope of this book and we will not examine it further. For more information regarding candlestick charting the reader can see Nison (1991) and references therein.

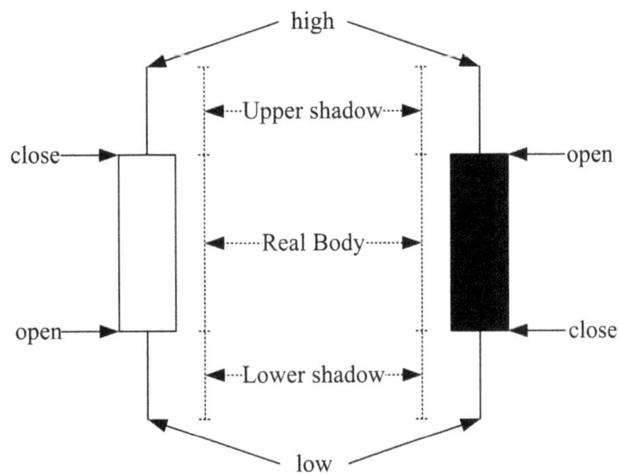

Fig. 1.2 Candlestick illustration

1.4.4 Filter Rules

Filter rules were initially proposed by Alexander (1961) with the purpose to identify and catch significant trends in financial assets' prices. A $x\%$ filter rule implies long positions after prices increase by $x\%$ and short positions when prices drop by $x\%$. Results reported in this study were against EMH, since filter rules were superior to buy-and-hold strategies when applied on Dow Jones and Standard & Poor's stock indices. Subsequent studies (Alexander 1964; Fama 1965; Fama and Blume 1966) illustrate that filter rules do not yield abnormal returns after considering transaction costs. Indications of excess returns did exist when small filters were applied, but as a consequence, these excess returns vanished after considering transaction costs resulted by the high frequency trading.

In Leuthold (1972) and Stevenson and Bear (1970), when filter rules applied on commodity futures they yield high rates of returns. However, Taylor (2008, pp. 200–201) notes that empirical results similar to those mentioned previously, cannot refute market efficiency for the examined commodity futures because there are three main problems to deal with (benchmarks, significance and optimization).

Although in earlier years filter rules were examined individually, as separate trading rules, in more recent surveys they consist parts of wider universes of trading rules tested (Falbo and Cristian 2011; Marshall et al. 2008a; Sullivan et al. 1999). Such studies usually test a set of numerous simple trading rules including, filter rules, moving averages, supports and resistances and channel breakouts.

However, filter rules are being also used as additional, necessary conditions for filtering and confirming signals of other technical trading rules, especially in the case of short term trading rules, that tend to generate numerous signals and thus high transaction costs. Examples of such short trading rules are short term moving averages, moving averages crossovers and trendline penetrations (Achelis 1995; Murphy 1986; Pring 2002). "False" trading signals usually reverse very quickly, and technicians use a variety of filters to eliminate these cases. Murphy (1986, pp. 242–245) presents the implementation of some of these filters on the case of moving averages. More precisely, he mentions that after an upwards (downwards) penetration of a moving average, some technicians require the whole subsequent day's price range to stay above (below) the average. Time filters are also common, in cases where a technician waits usually 1–3 days after the realization of a signal to verify its reliability.[6] Price filters are also used in order to characterise a penetration as valid. This means that price should breach a moving average by a fixed percentage (usually 1–3 %) to have an "accurate" signal. These filters, especially the time filters, can be also used in case of technical patterns, as useful tools to verify the confirmation of the pattern examined. These rules like most technical trading rules include some user-defined parameters. For example, in case of price filters, the size of the filter ($x\%$) is up to the discretional decision of the technician. The larger the filter, the more protection it provides. This means that more "*inaccurate*" signals are being omitted and less transaction costs we suffer. However in case of accurate signals, the implementation of a large filter sacrifices potential profits. Thus, the technician actually pays a protection-premium, which is losing a part of his/her potential future profits, due to his/her later market entry or exit. The inverse holds for small filter rules. It is thus an "*opportunity cost—opportunity benefit*" trade-off that the technician should consider within his trading strategy.

1.5 Controversial Perceptions for Technical Analysis

This section presents three main, longstanding, controversial perceptions for TA. Is TA science or art? Has TA any predictive power? If yes, is this attributed to the so called self-fulfilling prophesy? If not, is this due to TA's self-destructive nature? Since the performance of a particular technical trading rule depends on a number of user-defined parameters, how can someone find the values of the optimal combination of parameters to be used? Overfitting problems may result if back-testing techniques are used inappropriately for this purpose. This puzzling image of TA, the inability to pronounce it clearly and treat it as science and contrary this wide spread resonance to the academia and industry probably led Malkiel (1996, p. 139) to characterise TA as an "*anathema to the academic world...*" Generally, numerous are the critics against its usefulness. Characteristic is the book title of (Lo and Hasanhodzic 2009) "*The Heretics of Finance*", which is a package of conversations with leading practitioners of TA.

---

[6] Percentage envelopes (or volatility bands) and high-low bands are other types of filters described in (Murphy 1986) but their use cannot be generalized to other technical trading rules to the extent that filters, we have already presented, do.

1.5.1 Science Versus Art

Perhaps the most celebrated critique against TA is its subjective nature. This characterisation sets the most serious barrier so as to characterize TA as a science. In the context of technical patterns, subjectivity may refer to the fact that a particular price path could be differently interpreted by two experienced technicians. Hanley (2006) argues on that: "*Do we criticize medicine as being too subjective because two doctors have different interpretations of a set of symptoms? No, we simply conclude, based on outcomes, that perhaps one doctor is a better diagnostician than the other is*". But the subjectivity does not only exist in the interpretation but also in the identification of technical patterns. Via visual assessment on a particular price path, one technician might identify a pattern while another one might not. One of the main contributions of this book is the quantification of well-known technical patterns' recognition, in a manner that extracts this subjectivity.

A known debate found in the bibliography is the one of art vs. science. This debate is obvious in the three aforementioned definitions of TA (Sect. 1.2) and raised a lot of confusion in the academia. In Hanley (2006, p. 20) it is being cited that "*Art' means a skill acquired by experience, study or observation. 'Science' is a body of knowledge with its own axioms, rules, and language*". From our point of view, it is not appropriate to classify TA in any of the above two categories. It has common features with both terms, but also significant differences which does not allow us to proceed in such a distinction.

The following statement locates TA somewhere between art and science:

> It has been said that chart interpretation is not a science, but an art. It is not an exact science, to be sure, because it has no rules to which there are not exceptions. Its finer points defy expression in rule or precept. It requires judgment in appraisal of many factors, some of which may seem, at times, to conflict radically with others. But to call it an art, which implies the need for genius, or at least for a high degree of native talent, is certainly improper. Say, rather, that it demands skill, but a skill that can be acquired by anyone of ordinary intelligence.(Edwards and Magee 1997, p. 167)

Whether art or science, what is of prime importance is that TA is present in the daily investment decisions made by practitioners. Although there are no robust and persuasive empirical evidences suggesting that TA can accurately infer the direction of future prices diachronically, a substantial segment of the investment industry amalgamate TA with other trading and investment strategies. Taylor and Allen (1992) conducted a questionnaire survey on chief foreign exchange dealers regarding the use of TA. Among their findings, they reported that 90 % of respondents consider technical signals within their forecasts and investment decisions. The majority of those respondents combined the tools of fundamental and TA with a tendency of using the later for shorter horizons and fundamental analysis for longer ones. Their core results were corroborated for Germany in Menkhoff (1997), for Hong Kong in Lui and Mole (1998) and for USA in Cheung and Chinn (2001). Other studies that report similar findings concerning the usefulness of TA are

(Billingsley and Chance 1996; Fung and Hsieh 1997). Today there are numerous software programs and packages that deal with TA, whereas journals articles, newsletters and books are myriads. The presence of TA in the making decision process is apparent. What still remains a foggy issue to examine is its efficacy.

1.5.2 Self-Fulfilling Prophecy Versus Self-Destructive Nature

TA is also characterized as a "*self-fulfilling prophesy*" (Merton 1968). By this attribution, the predictive power that is assumed to be embedded in it, results by the persistence of technicians to believe in its efficacy even if this original conception is false. These initial common beliefs, guide technicians into homogenous investment actions, and thus force market prices to evolve as they had previously expected. In other words, their original false conceptions are verified, due to their massively common investment behaviour.[7] The term "*self-fulfilling prophecy*" is attributed to Robert Merton[8] (1968). More concretely he states:

> The self-fulfilling prophecy is, in the beginning, a false definition of the situation evoking a new behaviour which makes the original false conception come true. This specious validity of the self-fulfilling prophecy perpetuates a reign of error. For the prophet will cite the actual course of events as proof that he was right from the very beginning. (Merton 1968)

The fault of this argument as (Hanley 2006) documents is that in practice we cannot isolate, and observe the investment actions of technicians from the remaining investment community. Even if that would be plausible, the hypothesized homogeneity among technicians' interpretation of technical signals and investment actions is doubtful.

On the other side of the coin, TA is also characterized as "*self-destructive*" which contradicts with the former characterisation. According to this theory, a technical trading system is profitable until the time that it will be broadly used. After this point, it loses its predictability since market prices absorb this information (Timmermann and Granger 2004). In addition, a variety of other, more recent surveys suggest that trading profits from earlier profitable technical trading rules seem to vanish in more recent years (Olson 2004; Sullivan et al. 1999, 2003; Zapranis and Tsinaslanidis 2012b).

---

[7] The reader can see the survey-based articles of (Menkhoff 1997; Taylor and Allen 1992) where this attribution of TA is examined.

[8] The term "*self-fulfilling prophecy*" stems from the Thomas Theorem which was coined in 1928. According to this theorem "*if men define situations as real, they are real in their consequences*". For a comprehensive discussion on this theorem, see Merton (1995) and references therein.

1.5.3 Back-Testing Versus Overfitting

TA asserts that technical patterns can capture nonlinear regularities in financial price paths. A technician's task is to design technical trading rules each one of which is specified with a particular parameter combination. In the process of determining the optimal trading rule, technicians might use historical simulations (back-tests) which may result to overfitting. Assume that there is a deterministic nonlinear part in the examined dataset which can be captured by TA. However, real data exhibit a noisy part too. By overfitting, the discovered optimal trading rule captures both these parts. Since noise is not repeated into new data, the "optimal" trading rule does not actually perform optimally. The outcome of this process might be the identification of trading rules which perform best in-sample, but perform poorly out-of-sample, i.e. when new data are used. A technician should validate each trading rule in a validation sample, which is a subsample of the original dataset, not used in the optimal trading rule discovery process.

1.6 Subjective Nature and Behavioral Finance Critiques

The EMH has been for a long time a logical and consistent framework describing how market prices evolve. The majority of empirical evidences, which are in favour of it, are conductive to its dominant position in the academia. On the contrary, there are a sufficient large number of surveys challenging EMH. Minor or major in magnitude, market anomalies are apparent and this constitutes markets to be less than perfectly efficient. Since researches failed so far to fully explain the presence of such anomalies with models that assume rationality in investors' trading activities, a new field of economics known as behavioral finance provides further explanations accounting for psychological factors and irrationality.

Bibliography provides numerous types of cognitive biases which affect the manner investors act (Chincarini and Kim 2006; Haugen 1999; Shleifer 2000). In this section we are presenting only a selection of the biases mentioned in the bibliography. Behavioral finance can be used not only for purposes of providing explanations on the existence of market inefficiencies but also to answer the following question: If TA has no predictive power in a market, then why investors keep using it within their trading decisions? Behavioral finance can thus provide explanations for such irrationalities embedded in investors' decision making processes.

First, in laboratory experiments, subjects are often reported to have been searching for patterns in purely random sequences of stock prices. De Bondt (1998) noted that TA was similar to the fact that many investors see patterns where there are none. This is known as "*clustering illusion*". Gilovich (1993) states that the significance of this insight "*lies in the inescapable conclusion that our difficulty in accurately recognizing random arrangements of events can lead us to believe things that are not true and to believe something is systematic, ordered, and 'real' when it is really random, chaotic, and illusory"*. In addition he asserted: "*We are predisposed to see order, pattern, and meaning in the world; we find randomness, chaos, and meaninglessness unsatisfying. Human nature abhors a lack of predictability and the absence of meaning"*. The above statements may, at least partially, explain the general acceptance of technical patterns by many practitioners without even considering the efficacy of such patterns at all.

In a bundle of surveys, it is being reported that humans tend to be overconfident. *Overconfidence* implies that the majority of investors, thinks that is superior to the average investor, which is of course an irrational assertion. This bias may increase the frequency investors are trading, and thus transaction costs are also increased. Chincarini and Kim (2006) state that this bias is closely related with other two biases; *self-attribution* and *hind-sight* bias. According to the first one, a person attributes a success to his/her own skills (or talent) whereas he puts the blame to other people or conditions when a failure occurs. The *hind-sight* bias or else "*I-knew-it-all-along*" effect is the inclination to see past events as being predictable. This bias is closely linked with the *confirmation* bias which is the tendency to give emphasis on information that confirms rather than information that contradicts one's originally beliefs.

Another cognitive bias which violates the normative rules of investment behavior is the so-called "*neglect of probability*" bias. This bias captures the tendency of totally ignoring probabilities when making investment or trading decisions. It is worth to mention that whereas with *hind-sight* bias we actually misuse or misinterpret probabilities in decision making, with this bias we totally disregard probabilities when deciding.

Bibliography presents a bundle of surveys regarding the predictability of TA (see Sect. 1.4.2). Results are generally in favour of EMH. However an important number of published articles suggest that conditioning time periods and markets, specific patterns or technical indicators seemed to generate excess returns. But in general, no consistent profitability for specific technical tools is found, or at least published in the bibliography, for a relative large time period. Having this puzzling performance of TA, we cannot reasonably justify the great extent to which TA is being incorporated in investment decisions and the attention it gained throughout these years. The identification of technical patterns as described in technical books is too subjective, and the same stands for the interpretation of the returns generated by both technical patterns and indicators. Biases mentioned in this section may be used to explain partially this irrational, excess impact that TA is still having to academia and industry.

1.7 Purpose of the Book and Readership Level

The main purpose of this book is to resolve deficiencies and limitations that currently exist when using TA. Particularly, TA is being used either by academics as an "economic test" of the weak-form EMH or by practitioners as a main or supplementary tool for deriving trading signals. Existed bibliography which covers this dual use of TA usually suffers from critical problems. First, published empirical research use only a minor fraction of the available technical tools (mainly indicators) in order to evaluate the predictive behavior of TA and thus perform an economic test of the weak-form EMH. The subjective nature of TA enlarges the bias of such analyses especially when using technical patterns. Second, books on TA mainly deal with technical indicators, and when referring to patterns the approach followed is theoretical and descriptive rather than scientific and quantitative. In some cases only optimal examples are illustrated, which might give the false impression to readers, lacking the required scientific background, that charting is most of the times profitable. Statistical framework for assessing the realized returns is also usually absent. This book deals with the aforementioned problems by approaching TA in a systematic way. This is achieved through developing novel rule-based pattern recognizers, and the implementation of statistical tests for assessing the importance of their performance. Emphasis is given to technical patterns where subjectivity in their identification process is apparent. Our proposed methodology is based on the algorithmic and thus unbiased pattern recognition.

The methodological framework we present may prove to be useful for both future academic studies that test the null hypothesis of the weak-form EMH and for practitioners that want to embed TA within their trading/investment decision making processes. For brevity reasons, we selectively provide the corresponding Matlab scripts in chapters' appendices. However, the reader should not consider them as investment or trading recommendations. We strongly, encourage the reader, to modify the scripts, enhance them if necessary and construct his own trading strategy. The philosophy behind the design of the proposed algorithms, that were first developed in Tsinaslanidis (2012), is to capture the theoretical principles found in the literature for recognizing visually technical patterns and to quantify them accordingly.

1.8 Overview of the Book

Chapter 2 describes the necessary data preprocessing procedures that need to be followed before moving to the phase of technical pattern recognition. More precisely, the importance of detecting errors in a dataset and various manners of replacing missing values are discussed. In addition, a prerequisite step for the recognizing most technical patterns is the identification of sequences of regional locals. In this chapter two methodologies are presented for this purpose. The first one includes the use of a rolling window with a fixed size whilst the second utilizes perceptual important points.

Chapter 3 presents briefly some of the celebrated means by which someone can assess the predictive performance of TA. The statistical methods or measures presented here are classified into two general categories. The first one includes a bundle of methodologies that can be used to assess the performance trading signals derived by TA. The second category includes measures and methods that are used to assess predictions of models that use TA as an input variable.

In Chap. 4, we initially discuss various means by which horizontal support and resistance levels can be defined. Subsequently, we present our proposed algorithmic methodology which captures principles of identifying these levels via visual assessment (Zapranis and Tsinaslanidis 2012a). This algorithm is applied on NYSE and NASDAQ components, and its performance is subsequently assessed.

Chapter 5 focuses on a bundle of zigzag patterns that includes the head and shoulders, triple tops and bottoms, double tops and bottoms, flags, pennants and wedges. First the algorithmic recognition of these patterns is proposed. Subsequently, these recognizers are adopted on real and simulated series. Finally, the patterns' performance on both datasets is statistically compared.

Chapter 6 presents our proposed identification mechanism for the rounding bottoms and tops patterns (Zapranis and Tsinaslanidis 2012b). These patterns are considered to be circular and the proposed algorithm adopts simple geometric rules to capture the circular characteristics of these patterns. As for zigzag patterns, the proposed algorithm is adopted on real and simulated series and results obtained from both datasets are statistically assessed.

In Chap. 7 a bundle of well-known technical indicators is being presented. More precisely the corresponding mathematical formulas for their calculation, usual manners of interpreting and deriving from them trading signals are presented. The technical tools considered in this chapter are moving averages, moving average crossovers, moving average convergence/divergence, momentum, relative strength index, Bollinger bands and trading range breakouts.

Chapter 8 assesses the performance of trading signals generated from patterns examined in Chaps. 4–6 along with those derived from technical indicators presented in Chap. 7. The analysis followed is based on the analysis carried in Brock et al. (1992). More precisely, the performance of these technical trading rules is assessed by first adopting ordinary statistical tests, and second by following a bootstrap methodology.

Chapter 9 presents a Dynamic Time Warping (DTW) algorithmic process to identify similar patterns on a price series. This methodology initially became popular in applications of voice recognition, and it is not included in the context of TA. However it does rely on the same principles; that history is repeated, forming patterns which may vary in length. Patterns described in TA are numerous but finite, and our results (which are aligned with the majority of the empirical evidences provided in the literature) suggest that we should not expect a specific pattern to generate systematically abnormal returns over a long time period. In view of pattern recognition and finance applications, DTW can be used when the user wants to find in a given price series subsequences that are similar to a query series. For example, assume that the user defines this query to be the most recent part of a given series (say the last trading month). If the weak-form EMH does not hold, it may be possible to make predictions based on historical subsequences similar to this query.

References

Achelis SB (1995) Technical analysis from A to Z. Probus Publishing, Chicago

Alexander S (1961) Price movements in speculative markets: trends or random walks. Ind Manag Rev 2:7–26

Alexander S (1964) Price movements in speculative markets: trends or random walks, No 2. In: Cootner P (ed) The random character of stock market prices. MIT Press, Cambridge

Bachelier L (1900) Theory of speculation. In: Cootner P (ed) The random character of stock market prices. MIT press, Cambridge, MA, 1964; Reprint

Billingsley RS, Chance DM (1996) Benefits and limitations of diversification among commodity trading advisors. J Portf Manag 23:65–80

Bo L, Linyan S, Mweene R (2005) Empirical study of trading rule discovery in China stock market. Expert Syst Appl 28:531–535

Brock W, Lakonishok J, Lebaron B (1992) Simple technical trading rules and the stochastic properties of stock returns. J Finance 48:1731–1764

Bulkowski TN (2000) Encyclopedia of chart patterns. Wiley, New York

Campbell JY, Lo AW, MacKinlay AC (1997) The econometrics of financial markets. Princeton University Press, New Jersey

Chang PHK, Osler CL (1999) Methodical madness: technical analysis and the irrationality of exchange-rate forecasts. Econ J 109:636–661

Cheung Y, Chinn M (2001) Currency traders and exchange rate dynamics: a survey of the US markets. J Int Money Financ 20:439–471

Cheung W, Lam KSK, Yeung H (2011) Intertemporal profitability and the stability of technical analysis: evidences from the Hong Kong stock exchange. Appl Econ 43:1945–1963

Chincarini LB, Kim D (2006) Quantitative equity portfolio management: an active approach to portfolio construction and management. McGraw-Hill, New York

Clyde WC, Osler CL (1997) Charting: chaos theory in disguise? J Futur Mark 17:489–514

Coutts JA, Cheung K-C (2000) Trading rules and stock returns: some preliminary short run evidence from the Hang Seng 1985–1997. Appl Financ Econ 10:579–586

Cowles A (1933) Can stock market forecasters forecast? Econometrica 1:309–324

Curcio R, Goodhart C, Guillaume D, Payne R (1997) Do technical trading rules generate profits? Conclusions from the intra-day foreign exchange market. Int J Financ Econ 2:267–280

Dawson ER, Steeley JM (2003) On the existence of visual technical patterns in the UK stock market. J Bus Financ Account 30:263–293

De Bondt W (1998) A portrait of the individual investor. Eur Econ Rev 42:831–844

De Zwart G, Markwat T, Swinkels L, van Dijk D (2009) The economic value of fundamental and technical information in emerging currency markets. J Int Money Financ 28:581–604

Duda R, Hart P (1973) Pattern classification and scene analysis. Wiley, New York

Edwards RD, Magee J (1997) Technical analysis of stock trends, 7th edn. John Magee, Boston

Falbo P, Cristian P (2011) Stable classes of technical trading rules. Appl Econ 43:1769–1785

Fama E (1965) The behavior of stock market prices. J Bus 38:34–105

Fama E (1970) Efficient capital markets: a review of theory and empirical work. J Financ 25:383–417

Fama E (1991) Efficient capital markets II. J Financ 46:1575–1617

Fama E, Blume M (1966) Filter rules and stock-market trading. J Bus 39:226–241
Friesen GC, Weller PA, Dunham LM (2009) Price trends and patterns in technical analysis: a theoretical and empirical examination. J Bank Financ 33:1089–1100
Fung W, Hsieh DA (1997) Survivorship bias and investment style in the returns of CTAs. J Portf Manag 24:30–41
Gilovich T (1993) How we know what isn't so. Free Press, New York
Hamilton WP (1922) The stock market barometer. Harper Brothers, New York
Hanley PK (2006) Scientific frontiers and technical analysis. J Tech Anal 64:20–33
Hansen PR (2005) A test for superior predictive ability. J Bus Econ Stat 23:365–380
Haugen R (1999) The inefficient stock market. Prentice Hall, Upper Saddle River
Hsu P-H, Kuan C-M (2005) Reexamining the profitability of technical analysis with data snooping checks. J Financ Econ 3:606–628
Hudson R, Dempsey M, Keasey K (1996) A note on weak form efficiency of capital markets: the application of simple technical trading rules to UK prices 1935–1994. J Bank Financ 20:1121–1132
Leigh W, Modani N, Purvis R, Roberts T (2002a) Stock market trading rule discovery using technical charting heuristics. Expert Syst Appl 23:155–159
Leigh W, Paz N, Purvis R (2002b) Market timing: a test of a charting heuristic. Econ Lett 77:55–63
Leigh W, Purvis R, Ragusa JM (2002c) Forecasting the NYSE composite index with technical analysis, pattern recognizer, neural network, and genetic algorithm: a case study in romantic decision support. Decis Support Syst 32:361–377
Leigh W, Modani N, Hightower R (2004) A computational implementation of stock charting: abrupt volume increase as signal for movement in New York Stock exchange composite index. Decis Support Syst 37:515–530
Leigh W, Frohlich CJ, Hornik S, Purvis R, Roberts TL (2008) Trading with a stock chart heuristic. IEEE Trans Syst Man Cybern A Syst Hum 38:93–104
Leuthold RM (1972) Random walks and price trends: the live cattle futures markets. J Financ 27:879–889
Levy RA (1971) The predictive significance of five-point chart patterns. J Bus 44:316–323
Lo AW, Hasanhodzic J (2009) The heretics of finance: conversations with leading practitioners of technical analysis. Bloomberg Press, New York
Lo AW, Mamaysky H, Wang J (2000) Foundations of technical analysis: computational algorithms, statistical inference, and empirical implementation. J Financ 55:1705–1765
Lucke B (2003) Are technical trading rules profitable? Evidence for head-and-shoulder rules. Appl Econ 35:33–40
Lui Y, Mole D (1998) The use of fundamental and technical analysis by foreign exchange dealers: Hong Kong evidence. J Int Money Financ 17:535–545
Malkiel B (1996) A random walk down wall street. W.W. Norton & Company, New York
Mandelbrot B (1966) Forecasts of future prices, unbiased markets, and martingale models. J Bus 39:242–255
Marshall BR, Cahan RH, Cahan JM (2008a) Can commodity futures be profitably traded with quantitative market timing strategies? J Bank Financ 32:1810–1819
Marshall BR, Cahan RH, Cahan JM (2008b) Does intraday technical analysis in the U.S. equity market have value? J Empir Financ 15:199–210
Marshall BR, Qian S, Young M (2009) Is technical analysis profitable on US stocks with certain size, liquidity or industry characteristics? Appl Financ Econ 19:1213–1221
Menkhoff L (1997) Examining the use of technical currency analysis. Int J Financ Econ 2:307–318
Merton RK (1968) Social theory and social structure. Free Press, New York
Merton RK (1995) The Thomas theorem and the Matthew effect. Soc Forces 74:379–424
Mizrach B, Weerts S (2009) Highs and lows: a behavioural and technical analysis. Appl Financ Econ 19:767–777
Murphy J (1986) Technical analysis of the future markets: a comprehensive guide to trading methods and applications. Prentice Hall, New York

Neftci SN (1991) Naïve trading rules in financial markets and Wiener-Kolmogorov prediction theory: a study of "Technical Analysis" vol 64, pp. 549–71. J Bus 64:549–571
Nison S (1991) Japanese candlestick charting techniques. New York Institute of Finance, New York
Olson D (2004) Have trading rule profits in the currency markets declined over time? J Bank Financ 28:85–105
Osler CL (1998) Identifying noise traders: the head-and-shoulders pattern in U.S. equities. Federal Reserve Bank of New York Staff Reports (Vol. 42)
Osler CL (2000) Support for resistance: technical analysis and intraday exchange rates. FRBNY Economic Policy Review 6(2):53–68
Osler CL (2002) Stop-loss orders and price cascades in currency markets. Mimeo, Federal Reserve Bank of New York, New York
Osler CL (2003) Currency orders and exchange rate dynamics: an explanation for the predictive success of technical analysis. J Financ 58:1791–1819
Osler CL (2005) Stop-loss orders and price cascades in currency markets. J Int Money Financ 24:219–241
Osler CL, Chang PHK (1995) Head and shoulders: not just a flaky pattern. Federal Reserve Bank of New York Staff Reports (Vol. 4)
Pring MJ (2002) Technical analysis explained: the successful investor's guide to spotting investment trends and turning points, 4th edn. McGraw-Hill, New York
Qi M, Wu Y (2006) Technical trading-rule profitability, data snooping, and reality check: evidence from the foreign exchange market. J Money Credit Bank 38:2135–2158
Raj M, Thurston D (1996) Effectiveness of simple technical trading rules in the Hong Kong futures markets. Appl Econ Lett 3:33–36
Rhea R (1932) Dow theory. Barron's, New York
Roberts H (1967) Statistical versus clinical prediction of the stock market. Unpublished manuscript, Center for Research in Security Prices, University of Chicago
Samuelson P (1965) Proof that properly anticipated prices fluctuate randomly. Ind Manag Rev 6:41–49
Savin G, Weller P, Zvingelis J (2007) The predictive power of "head-and-shoulders" price patterns in the U.S. stock market. J Financ Econ 5:243–265
Shleifer A (2000) Inefficient markets: an introduction to behavioral finance. Oxford University Press, New York
Stevenson RA, Bear RM (1970) Commodity futures: trends or random walks? J Financ 25:65–81
Sullivan R, Timmermann A, White H (1999) Data-snooping, technical trading rule performance, and the bootstrap. J Financ LIV:1647–1691
Sullivan R, Timmermann A, White H (2003) Forecast evaluation with shared data sets. Int J Forecast 19:217–227
Taylor SJ (2008) Modelling financial time series, 2nd edn. World Scientific, Singapore
Taylor MP, Allen H (1992) The use of technical analysis in the foreign exchange market. J Int Money Financ 11:304–314
Timmermann A, Granger C (2004) Efficient market hypothesis and forecasting. Int J Forecast 20:15–27
Tsinaslanidis P (2012) Technical trading strategies, pattern recognition and weak-form market efficiency tests. University of Macedonia, Thessaloniki
Wang J, Chan S (2007) Stock market trading rule discovery using pattern recognition and technical analysis. Expert Syst Appl 33:304–315
Wang J, Chan S (2009) Trading rule discovery in the US stock market: an empirical study. Expert Syst Appl 36:5450–5455
White H (2000) A reality check for data snooping. Econometrica 68:1097–1126
Zapranis A, Tsinaslanidis P (2010a) A behavioral view of the head-and-shoulders technical analysis pattern. In: Proceedings of the 3rd international conference in accounting and finance, Skiathos, Greece, pp 1515–1529

Zapranis A, Tsinaslanidis P (2010b) Identification of the head-and-shoulders technical analysis pattern with neural networks. In: Artificial neural networks—ICANN 2010, 20th international conference, vol 6354. Springer, Thessaloniki, Greece, pp 130–136)
Zapranis A, Tsinaslanidis PE (2012a) Identifying and evaluating horizontal support and resistance levels: an empirical study on US stock markets. Appl Financ Econ 22:1571–1585
Zapranis A, Tsinaslanidis PE (2012b) A novel, rule-based technical pattern identification mechanism: identifying and evaluating saucers and resistant levels in the US stock market. Expert Syst Appl 39:6301–6308

Chapter 2
Preprocessing Procedures

2.1 Introduction

The purpose of this chapter is to present two important preprocessing procedures than need to be carried before someone moves to the phase of recognizing technical patterns in financial price series. First, the importance of detecting errors in a dataset (the editing process) and various manners of replacing missing values (imputation) are discussed. Second, after ensuring that datasets are cleaned, two methodologies of identifying regional peaks and bottoms are presented. The first method presented is the identification of regional locals by using a rolling window of fixed size, while the second method includes the identification of local extrema known in the context of data mining as perceptually important points. The identification of these regional locals is crucial in technical pattern recognition processes since the criteria used for identifying a pattern mainly refer to sequences of local extrema.

The rest of the chapter is organized as follows. In Sect. 2.2, means of dealing with problematic datasets are discussed. In Sect. 2.3 two methods for the identification of regional locals in price series are presented. More precisely, the use of a rolling window is presented in Sect. 2.3.1, while the identification of perceptually important points is presented in Sect. 2.3.2. Finally, in Sect. 2.4 we conclude. In Appendices 1 and 2, the reader will find the developed Matlab code for the methodology presented in Sects. 2.3.1 and 2.3.2 respectively.

2.2 Data Pre-processing

Even high quality data purchased from well reputed vendors might be problematic. The quality of data used is crucial in any empirical assessment. If not carefully pre-processed false conclusions might be reached, good trading systems

P.E. Tsinaslanidis, A.D. Zapranis, *Technical Analysis for Algorithmic Pattern Recognition*, DOI 10.1007/978-3-319-23636-0_2

might be rejected whilst in a worst-case scenario bad trading systems might be approved. The old acronym GIGO mainly used in the field of computer science can also be used in the scientific field of finance. It stands for "Garbage In, Garbage Out", and means that if invalid input data are used, invalid outputs will result.

The process of detecting errors in statistical data is called editing. Chambers (2001) mentions two different editing types: logical and statistical editing. The first one refers to the process of detecting data values which violate particular pre-defined rules. Values that fail a logical edit must be wrong and we call them illogical. The second type of editing is the process of identifying suspicious values, i.e. values that might be wrong. Values that fail the editing process are initially replaced as missing values and subsequently if necessary are replaced by known acceptable values, a process called imputation. The necessity of imputation varies, but generally it is being adopted if the trading system considered is designed to work with complete datasets.

Illogical values, as their name suggests, violate logical constraints. Common examples of illogical values are observing negative values for the trading volume, data falling on days where the market is closed (weekends or holidays) and bad quotes where for example a tick of 21.62 should be 31.62. Detecting illogical values can be in most cases a straightforward process. Consider the following example where the variable of interest is the market price of a stock. By calculating daily returns, we can easily spot daily returns which exceed the maximum decline/rise permitted, i.e. they violate the limit up or limit down boundaries (e.g. a daily change of 54 %). Such cases might result from missing prices or bad quotes as well. More generally, distances of high, low and close prices from the corresponding open prices can also be used for identifying illogical values.

Outliers are extreme values, whose presence are theoretically possible, but raise suspicions. One way to spot such cases is by calculating the variable's mean and the standard deviation, and subsequently identify observations distanced more than ± 3 standard deviations from the mean. After a careful examination, if we conclude that an outlier is an error then we initially record it as a missing value. Kumiega and Van Vliet (2008) describe another data cleaning process for outliers. By this process, a compressing algorithm winsorizes outliers, by pulling them towards the mean and replacing them with a value at specified limit, say three standard deviations. However, in view of finance applications, where financial asset's returns exhibit time-varying volatility Kumiega and Van Vliet (2008) recommend winsorizing on a rolling basis.

Either because there have been missing values in the initial dataset, or because we replaced values failing the editing process with missing ones, the outcome is a dataset with "holes". Generally, there are two options to deal with this problem. We can either ignore these missing values by not including them in our final dataset, or we can replace them with new acceptable values. The final choice depends on the application considered. The first procedure has the drawback of losing probable useful information from the initial sample. In addition, when the number of errors is significant, ignoring missing values will reduce the sample size. For the second alternative two methodologies can be used. In the first methodology, called donor imputation, the missing values for one or more variables, called recipients, are replaced by the corresponding similar to them values called donors (Beaumont and Bocci 2009). If donors are more than one then the replacing value is determined by various ways like choosing the first candidate donor, calculating the average value of all donors, choosing randomly among a set of potential donors (Random Hot-Deck imputation) or using a Nearest-Neighbour approach (Nearest-Neighbour) imputation. Alternatively, missing values can be replaced by acceptable values by using a moving average or a more complex model (like an ARIMA[1]) prior to the missing value. Finally, missing values of financial price series can also be replaced by linearly interpolating adjacent values.

However, cleaning data might also raise some problems. It is not certain whether the cleaned data were the values observed, when the actual trading took place in real time (Kumiega and Van Vliet 2008). Thus, cleaned data, when used for back-testing a trading system, might affect the results obtained and their inferences.

2.3 Identification of Regional Locals

Financial price series have idiosyncratic characteristics over other price series. Significant points (like regional locals and turning points) are crucial in pattern recognition processes via visual assessment. Descriptions found in the literature, for the visual identification of technical patterns, involve the assignment of criteria (conditions) to sequences of regional locals. More precisely, Neftci (1991) states:

> ...most patterns used by technical analysts need to be characterized by appropriate sequences of local minima and/or maxima...

This section presents two methodologies for identifying regional locals. More precisely, in Sect. 2.3.1 the identification of local minima and maxima with a rolling window is described while in Sect. 2.3.2 the identification of perceptually important points is presented.

[1] Autoregressive Integrated Moving Average.

2.3.1 Identify Regional Locals with a Rolling Window

The prerequisite process for the identification of the technical patterns examined in this book, involves the identification of regional peaks and bottoms with the following developed Matlab function (see Appendix 1)[2]:

$$[Peaks, Bottoms] = RW(ys, w, pflag)$$

The $RW(\cdot)$ function takes three input variables: *ys*, *w* and *pflag*. "*ys*" is a column vector of the y-coordinates (prices) of the examined price series, "*w*" is used to define the width of the *rolling window* (RW), and "*pflag*" returns a corresponding graph if it takes the value of one. The outputs are the regional locals identified by the function. Particularly, ***Peaks*** (***Bottoms***) is a $m \times 2$ ($k \times 2$) matrix containing the coordinates of the m (k) identified peaks (bottoms). The first and second columns of the above outputs contain the y and x-coordinates respectively. For a given price series, an observation is identified as local peak (trough) if it is the largest (smallest) of the observations in a window of size $2w + 1$ centered on this observation. The window slides by one observation in each iteration and the process is repeated until the whole price series is scanned (Kugiumtzis et al. 2007).

Let $\{p_t\}_{t=1}^{\ell}$ and $t \in [1 : \ell]$ be the prices (y-coordinates) and the time (x-coordinates) of the examined price series of length ℓ respectively.[3] The indicator t refers to the oldest observation when it takes the value of one and to the most recent observation when it takes the value of ℓ respectively. The process for identifying the regional locals is described in (2.1) and (2.2) $\forall\, t \in [w + 1 : \ell - w]$.

$$Local\ Peak\ if\ \ p_t > \max\left\{p_{[t-w:t-1]}\right\} \&\ p_t > \max\left\{p_{[t+1:t+w]}\right\} \tag{2.1}$$

$$Local\ Trough\ if\ \ p_t < \min\left\{p_{[t-w:t-1]}\right\} \&\ p_t < \min\left\{p_{[t+1:t+w]}\right\} \tag{2.2}$$

An example is illustrated in Fig. 2.1 where $RW(\cdot)$ identified 6 regional peaks and seven regional bottoms (troughs) on NASDAQ Index by adopting a rolling window of 31 days ($w = 15$). The identification process starts from the 16th observation ($t = 16$) and terminates when $t = \ell - 15$, where ℓ is the length of the price series.

Alternative methods to identify regional locals are also provided in the bibliography. By implementing kernel mean regression algorithm we can smooth the price series and identify the corresponding extrema (Lo et al. 2000; Dawson and Steeley

---

[2] Functions presented in this book have the following general form: [output_1, output_2, …, output_n] = function's name (input_1, input_2, …, input_n). The variables in the squared brackets are the outputs generated by the corresponding function and the variables inside the brackets are the necessary inputs. We follow the same notation used in the Matlab software since all the identification mechanisms presented in this book were developed with the use of this software.

[3] Hereafter we will use the notation $[a : b]$ to refer to all positive natural values between the closed interval $[a, b]$, where $0 < a < b$ and $a, b \in \mathbb{N}$.

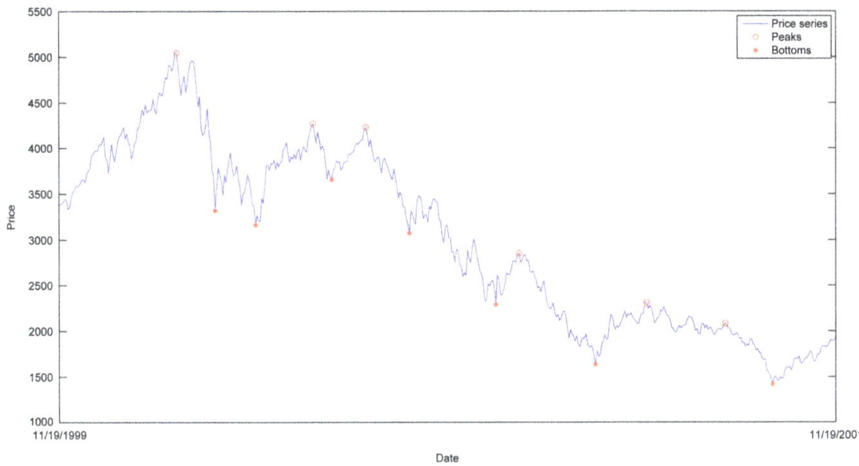

Fig. 2.1 Identification of regional locals on NASDAQ Index. Six Peaks and seven Bottoms were identified by $RW(\cdot)$ with a rolling window of 31 days (w = 15)

2003; Savin et al. 2007). Lucke (2003) used a computer program which was originally designed to identify business cycle turning points. Identification of perceptually important points is an alternative algorithmic approach for detecting regional significant points and it is presented in Sect. 2.3.2.

2.3.2 Perceptually Important Points

A promising method to exploit salient points from a price series is by using *Perceptually Important Points* (PIPs). The identification methodology was first introduced in Chung et al. (2001), and used in many applications on time series data mining. More precisely, they have been used mainly for purposes of dimension reduction (or else time series representation) (Fu et al. 2008; Phetchanchai et al. 2010), as a dynamic approach for time series segmentation (Fu et al. 2006; Jiang et al. 2007; Tsinaslanidis and Kugiumtzis 2014) and for clustering reasons (Fu et al. 2004).[4] PIPs have been also used in finance applications to identify technical patterns (Fu et al. 2007; Chen et al. 2013).

As a preprocessing step of our methodology, we may use PIPs in order to identify significant points. The algorithm starts by characterizing the first and the last observation as the first two PIPs. Subsequently, it calculates the distance between all remaining observations and the two initial PIPs and signifies as the third PIP the one with the maximum distance. The fourth PIP is the point that

[4] A comprehensive review on the existing time series data mining research is presented in Fu (2011), where variant methodologies that deal with the aforementioned aspects of data mining are highlighted.

maximizes its distance to its adjacent PIPs (which are either the first and the third, or the third and the second PIP). The algorithm stops when the required by the user number of PIPs is identified.

Three metrics are generally used for the distance in the PIPs algorithm, namely the Euclidean distance (ED) d_E, the perpendicular distance (PD) d_P and the vertical distance (VD) d_V. Let $\{p_1, p_2, \ldots, p_\ell\}$ be the price time series of length ℓ, and two adjacent PIPs $x_t = (t, p_t)$ and $x_{t+T} = \left(t+T, p_{t+T}\right)$. The Euclidean distance d_E of each of the intermediate points $x_i = (i, p_i)$, for $i \in \{t+1, \ldots, t+T-1\}$ from the two PIPs is defined in (2.3).

$$d_E(x_i, x_t, x_{t+T}) = \sqrt{(t-i)^2 + (p_t - p_i)^2} + \sqrt{(t+T-i)^2 + \left(p_{t+T} - p_i\right)^2} \tag{2.3}$$

For the two other distances, we consider first the line connecting the two PIPs $x_t = (t, p_t)$ and $x_{t+T} = \left(t+T, p_{t+T}\right)$, $z_i = s\,i + c$, and (i, z_i) the points on the line, where the slope is $s = \left(p_{t+T} - p_t\right)/T$ and the constant term is $c = p_t - t\left(p_{t+T} - p_t\right)/T$. Then the perpendicular distance d_P of any intermediate point $x_i = (i, p_i)$, between the two PIPs from the line is given by (2.4).

$$d_P(x_i, x_t, x_{t+T}) = \frac{|s\,i + c - p_i|}{\sqrt{s^2+1}} \tag{2.4}$$

Finally, (2.5) expresses the vertical distance d_V of x_i to the line.

$$d_V(x_i, x_t, x_{t+T}) = |s\,i + c - p_i| \tag{2.5}$$

For any of the three distances, denoted collectively d, the new PIP point, $x_i^* = \left(i^*, p_{i^*}\right)$, is the one that maximizes the distance d at i^* (2.6). In (2.6) "*argmax*" stands for the argument of maximum.

$$i^* = \underset{i}{\operatorname{argmax}} \left(d(x_i, x_t, x_{t+T})\right) \tag{2.6}$$

Figure 2.2 presents a step-by-step identification process of five ED-PIPs on NASDAQ Index for the requested period 19/11/1999–19/11/2001. Figure 2.3 presents a simplified illustration of the identification of the third PIP according with the three aforementioned distance measures.

At this point the algorithmic methodology of PIPs' identification is presented. Let $\boldsymbol{P} = \{p_t\}_{t=1}^{\ell}$ and $\mathbf{T} = \{t\}_{t=1}^{\ell}$ are two $\ell \times 1$ column vectors containing the prices (y-coordinates) and the time (x-coordinates) of the examined price series of length ℓ respectively. We subsequently define $\boldsymbol{A_{x\imath}}$ and $\boldsymbol{A_{y\imath}}$ two $\ell \times 2$ matrices containing the x- and y-coordinates respectively, of the closer adjacent PIPs for the $\imath^{th}$ iteration.[5]

[5] Since the first two PIPs are defined as the first and the last observation the $\imath^{th}$ iteration identifies the $(\imath+2)^{th}$ PIP.

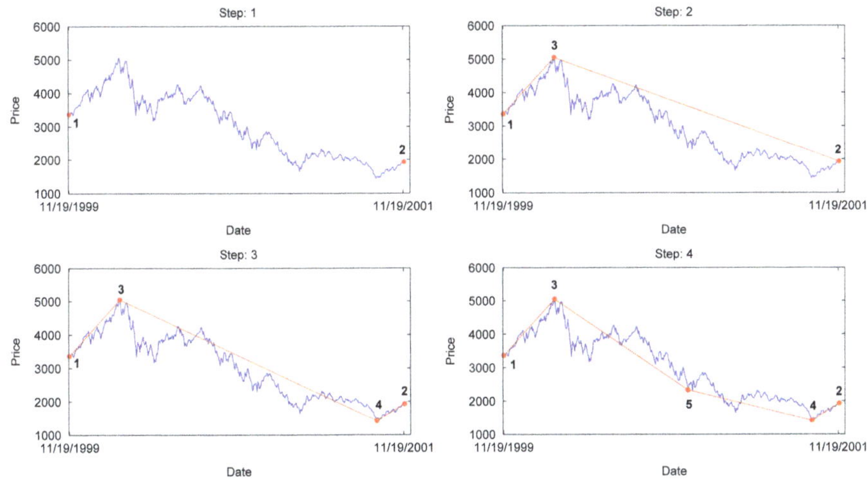

Fig. 2.2 PIPs identification process. First 5 PIPs identified on NASDAQ Index, with the ED measure

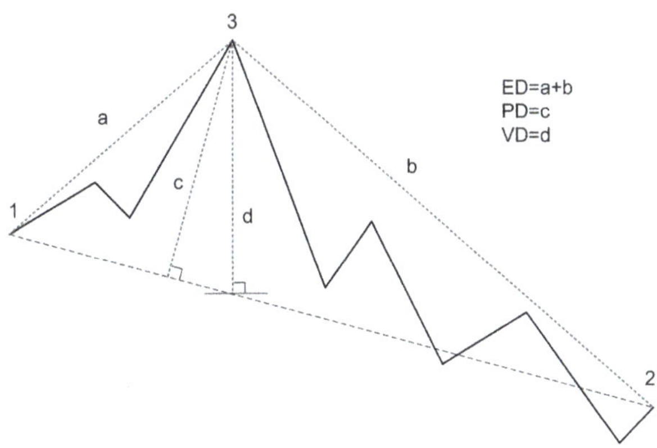

Fig. 2.3 Identification of the third PIP with three different distance measures ED, PD and VD

Particularly, the first column contains the coordinates of the closer adjacent PIP whereas the second column contains the coordinates of the second closer adjacent PIP. Existed PIPs are indicated with "*nan*"[6] in order to avoid the identification of the same PIP. Consider the following simplified example. If $\ell = 6$, in the first iteration, $\imath = 1$, the third PIP is about to identified (since the first two PIPs are $(1, p_1)$ and $(6, p_6)$), and $\boldsymbol{A}_{x1}$ and $\boldsymbol{A}_{y1}$ are shown in (2.7) and (2.8).

[6] *nan* stands for "not-a-number". A nan value is the result from operations which have undefined numerical results. When nan is involved in a calculation (for example $nan \times 10$) the result is also nan.

$$\boldsymbol{A}_{x1} = \begin{bmatrix} nan & 1 & 1 & 6 & 6 & nan \\ nan & 6 & 6 & 1 & 1 & nan \end{bmatrix}' \tag{2.7}$$

$$\boldsymbol{A}_{y1} = \begin{bmatrix} nan & p_1 & p_1 & p_6 & p_6 & nan \\ nan & p_6 & p_6 & p_1 & p_1 & nan \end{bmatrix}' \tag{2.8}$$

In (2.7) and (2.8), $\boldsymbol{X}'$ is the transpose of $\boldsymbol{X}$. If the third identified PIP is $(4, p_4)$, in the subsequent iteration, $\imath = 2$, (2.7) becomes

$$\boldsymbol{A}_{x2} = \begin{bmatrix} nan & 1 & 4 & nan & 4 & nan \\ nan & 4 & 1 & nan & 6 & nan \end{bmatrix}' \tag{2.9}$$

and (2.8)

$$\boldsymbol{A}_{y2} = \begin{bmatrix} nan & p_1 & p_4 & nan & p_4 & nan \\ nan & p_4 & p_1 & nan & p_6 & nan \end{bmatrix}'. \tag{2.10}$$

In every iteration, (2.11) measures the ED between the adjacent PIPs and intermediate points.

$$\begin{aligned} \boldsymbol{ED}_\imath = & \left\{ [\boldsymbol{A}_{x\imath}\,(:,1) - \boldsymbol{T}]^{\circ 2} + \left[\boldsymbol{A}_{y\imath}\,(:,1) - \mathbf{P}\right]^{\circ 2} \right\}^{\circ 1/2} \\ & + \left\{ [\boldsymbol{A}_{x\imath}\,(:,2) - \boldsymbol{T}]^{\circ 2} + \left[\boldsymbol{A}_{y\imath}(:,2) - \mathbf{P}\right]^{\circ 2} \right\}^{\circ 1/2} \end{aligned} \tag{2.11}$$

Here, $\boldsymbol{ED}_\imath = \{ed_{t,\imath}\}_{t=1}^{\ell}$ is an $\ell \times 1$ column vector containing the Euclidean distances of all intermediate points from their two adjacent PIPs for the $\imath^{th}$ iteration, $\circ$ symbolizes the Hadamard (or else element-wise) product, $\boldsymbol{X}^{\circ n}$ is the nth element-wise power of the matrix $\boldsymbol{X}, n \in \mathbb{R}$, and $\boldsymbol{X}(:,j)$ represents the jth column of matrix $\boldsymbol{X}$. The new PIP identified in $\imath^{th}$ iteration (2.12) has coordinates $PIP_\imath = \left(PIP_{x,\imath}, PIP_{y,\imath}\right)$, where

$$PIP_\imath = \left(\underset{t}{\text{argmax}}(ed_{t,\imath}),\, p_{PIP_{x,\imath}} \right). \tag{2.12}$$

Alternatively, we can identify PIPs by measuring Perpendicular Distances (PD) of the intermediate points from the lines passing through their two adjacent PIPs. As already mentioned, on a two-dimensional, Cartesian coordinate system, the points (i, z_i) of a straight line are defined by $z_i = s\,i + c$, where s is the slope of the line and c is the constant term. Equations (2.13) and (2.14) show the calculation of $\boldsymbol{S}_\imath$ and $\boldsymbol{C}_\imath$ respectively, which are two $\ell \times 1$ column vectors containing accordingly the slopes and constant terms of lines defined by all successive pairs of existed PIPs, identified before the completion of the ιth iteration.

$$\boldsymbol{S}_\iota = \frac{\boldsymbol{A}_{y\iota}(:,1) - \boldsymbol{A}_{y\iota}(:,2)}{\boldsymbol{A}_{x\iota}(:,1) - \boldsymbol{A}_{x\iota}(:,2)} \tag{2.13}$$

$$\boldsymbol{C}_\iota = \boldsymbol{A}_{y\iota}(:,1) - \boldsymbol{S}_\iota \circ \boldsymbol{A}_{x\iota}(:,1) = \boldsymbol{A}_{y\iota}(:,2) - \boldsymbol{S}_\iota \circ \boldsymbol{A}_{x\iota}(:,2) \tag{2.14}$$

Subsequently, $\boldsymbol{PD}_\iota = \{pd_{t,\iota}\}_{t=1}^{\ell}$ is a $\ell \times 1$ column vector containing all points' corresponding perpendicular distances as measured within the ιth iteration (2.15).[7]

$$\boldsymbol{PD}_\iota = \frac{|\boldsymbol{S}_\iota \circ \boldsymbol{T} - \boldsymbol{P} + \boldsymbol{C}_\iota|}{(\boldsymbol{S}_\iota^{\circ 2} + 1)^{\circ 1/2}} \tag{2.15}$$

Similarly with the case of ED-PIPs the new PD-PIP has coordinates

$$PIP_\iota = \left(\underset{t}{\operatorname{argmax}}(pd_{t,\iota}),\, p_{PIP_{x,\iota}}\right). \tag{2.16}$$

Finally by using the Vertical Distance (VD) as a distance measure, the $\boldsymbol{VD}_\iota = \{vd_{t,\iota}\}_{t=1}^{\ell}$ is an $\ell \times 1$ column vector defined by (2.17)

$$\boldsymbol{VD}_\iota = |\boldsymbol{Z}_\iota - \boldsymbol{P}| \tag{2.17}$$

where

$$\boldsymbol{Z}_\iota = \boldsymbol{S}_\iota \circ \mathbf{T} + \boldsymbol{C}_\iota. \tag{2.18}$$

Similarly, the new VD-PIP has coordinates

$$PIP_\iota = \left(\underset{t}{\operatorname{argmax}}(vd_{t,\iota}),\, p_{PIP_{x,\iota}}\right) \tag{2.19}$$

We prefer to use matrix notations for the above calculations in order to avoid "*for-loops*" in programming and enhance the computational speed. The Matlab code for the PIPs identification process we described above is provided in Appendix 2. Alternatively the reader may use the relevant equations referred in Fu et al. (2008) or (2.3)–(2.6).

---

[7] Since the division between matrices is not defined in Eqs. (2.13) and (2.14) an element wise division is implied.

2.4 Conclusions

For reliable results when assessing a trading system the use of high-quality data is necessary. Data retrieved even from well-reputed vendors may be problematic. Illogical, suspicious and missing values are the most common problems that need to be dealt before performing an empirical analysis. In this chapter these issues were discussed and various methods for dealing with these problems were highlighted.

The second part of this chapter focuses on the identification of regional locals on financial price series. Someone who is already familiar with the descriptions for identifying various technical patterns, should have noticed that these descriptions mainly refer to specific criteria that a sequence of regional locals must fulfil for a successful pattern confirmation. In this chapter two methodologies for identifying regional locals were presented. The first includes the use of a RW centered on each observation. This observation is characterized as a local peak (bottom) if it is the maximum (minimum) of all observations included in this window. This is a sequential algorithmic approach which starts from the first and terminates to the last available observation. This means that locals that correspond to earlier observations are identified first. When using RW the user must set the window size and the number of regional locals that will be identified is not known, a priori. On the contrary, PIPs identification is an algorithmic process where the user must specify the distance measure to be used and the number of regional locals to be identified. In this process a series is scanned dynamically until the desired number of regional locals is identified. It is up to the discretion of the user to decide which one of the presented methods (if any) should be used.

Appendix 1: RW Function

1.1. The Function

$$[Peaks, Bottoms] = RW(ys, w, pflag)$$

1.2. Description

This function identifies regional peaks and bottoms of a price series with a RW of size $2w + 1$, and a slide step of one observation.

Inputs

1. *ys*: Price series (Column Vector).
2. *w*: Is used to define the size of the RW, which is $2w + 1$.
3. *pflag*: If 1, the function generates a plot.

Outputs

1. *Peaks*: An $(n \times 2)$ matrix containing the coordinates of the n identified peaks.
2. *Bottoms*: A $(k \times 2)$ matrix containing the coordinates of the k identified bottoms.

   The first and second column of these outputs contains the *y*- and *x*-coordinates respectively.

1.3. Code

```
function [Peaks,Bottoms]=RW(ys,w,pflag)
l=length(ys);
Peaks_Bottoms=zeros(l,2);%Preallocation
for i=w+1:l-w%Index peaks and bottoms with ones
  if ys(i,1)>max(ys(i-w:i-1))...
      && ys(i,1)>max(ys(i+1:i+w))
    Peaks_Bottoms(i,1)=1;
  end
  if ys(i,1)<min(ys(i-w:i-1))...
      && ys(i,1)<min(ys(i+1:i+w))
    Peaks_Bottoms(i,2)=1;
  end
end
P_Indx=find(Peaks_Bottoms(:,1));
B_Indx=find(Peaks_Bottoms(:,2));
Peaks=[ys(P_Indx),P_Indx];
Bottoms=[ys(B_Indx),B_Indx];
if pflag==1
  plot(ys),hold on
  plot(Peaks(:,2),Peaks(:,1),'ro')
  plot(Bottoms(:,2),Bottoms(:,1),'r*')
  legend('Price series','Peaks','Bottoms'), hold off
end
end
```

Appendix 2: PIPs Function

2.1. The Function

$$[PIPxy] = PIPs(ys, n\ of\ PIPs, type\ of\ dist, pflag)$$

2.2. Description

This function identifies the Perceptually Important Points (PIPs) on a price series.

Inputs

1. *ys*: Price series (Column Vector).
2. *n of PIPS*: Number of requested PIPs.
3. *type of dist*: 1 = (Euclidean Distance) ED, 2 = (Perpendicular Distance) PD and 3 = (Vertical Distance) VD.
4. *pflag*: If 1, the function generates a plot.

Outputs

1. *PIPxy*: A $(n\,of\,PIPs \times 2)$ matrix containing the coordinates of PIPs. The first (second) column presents the x-coordinates (y-coordinates).

2.3. Code

```
function [PIPxy]=PIPs(ys,nofPIPs,typeofdist,pflag)
l=length(ys);
xs=(1:l)';% Column vector with xs

PIP_points=zeros(l,1);% Binary indexation
PIP_points([1,l],1)=1;% One indicate the PIP points.The first two
PIPs are the first and the last observation.

Adjacents=zeros(l,2);
currentstate=2;% Initial PIPs

while currentstate<=nofPIPs
  Existed_Pips=find(PIP_points);
  currentstate=length(Existed_Pips);
  locator=nan(l,currentstate);
  for j=1:currentstate
      locator(:,j)=abs(xs-Existed_Pips(j,1));
  end
  b1=zeros(l,1);b2=b1;
```

```
  for i=1:l
    [~,b1(i)]=min(locator(i,:));% Closer point
    locator(i,b1(i))=nan; % Do not consider Closer point
    [~,b2(i)]=min(locator(i,:));% 2nd Closer Point
         Adjacents(i,1)=Existed_Pips(b1(i));%x-coordinates of the
closer point
         Adjacents(i,2)=Existed_Pips(b2(i));%x-coordinates of the
2nd closer points
  end
  %% Calculate Distance
  Adjx=Adjacents;
  Adjy=[ys(Adjacents(:,1)),ys(Adjacents(:,2))];
    Adjx(Existed_Pips,:)=nan;% Existed PIPs are not candidates for
new PIP.
  Adjy(Existed_Pips,:)=nan;
  if typeofdist==1
    [D]=EDist(ys,xs,Adjx,Adjy);
  elseif typeofdist==2
    [D]=PDist(ys,xs,Adjx,Adjy);
  else
    [D]=VDist(ys,xs,Adjx,Adjy);
  end
  [~,Dmax]=max(D);
  PIP_points(Dmax,1)=1;
  currentstate=currentstate+1;
end
PIPxy=[Existed_Pips, ys(Existed_Pips)];
%% Plot
if pflag==1
  plot(ys), hold on
  plot(Existed_Pips,ys(Existed_Pips),'r*'),hold off
end
end
%% Distance measures
% Euclidean Distance
function [ED]=EDist(ys,xs,Adjx,Adjy)
ED=((Adjx(:,2)-xs).^2+(Adjy(:,2)-ys).^2).^(1/2)+...
  ((Adjx(:,1)-xs).^2+(Adjy(:,1)-ys).^2).^(1/2);
end
% Perpendicular Distance
function [PD]=PDist(ys,xs,Adjx,Adjy)
slopes=(Adjy(:,2)-Adjy(:,1))./(Adjx(:,2)-Adjx(:,1));
constants=Adjy(:,2)-slopes.*Adjx(:,2);
PD=abs(slopes.*xs-ys+constants)./(slopes.^2+1).^(1/2);
% line function: y=kx+m (1)
```

```
% the perpendicular distance (PD) from a point p(x1,y1) to a line
% is given by the following formula:
% PD=abs(k*x1-y1+m)/sqrt(k^2+1)
end
% Vertical Distance
function [VD]=VDist(ys,xs,Adjx,Adjy)
slopes=(Adjy(:,2)-Adjy(:,1))./(Adjx(:,2)-Adjx(:,1));
constants=Adjy(:,2)-slopes.*Adjx(:,2);
Yshat=slopes.*xs+constants;
VD=abs(Yshat-ys);
end
```

References

Beaumont J-F, Bocci C (2009) Variance estimation when donor imputation is used to fill in missing values. Can J Stat 37(3):400–416

Chambers R (2001) Evaluation criteria for statistical editing and imputation. National Statistics Methodological Series No. 28. Office for National Statistics, UK

Chen C-H, Tseng VS, Yu H-H, Hong T-P (2013) Time series pattern discovery by a PIP-based evolutionary approach. Soft Comput 17:1699–1710

Chung FL, Fu TC, Luk R, Ng V (2001) Flexible time series pattern matching based on perceptually important points. Paper presented at the international joint conference on artificial intelligence workshop on learning from temporal and spatial data

Dawson ER, Steeley JM (2003) On the existence of visual technical patterns in the UK stock market. J Bus Financ Account 30(1 and 2):263–293

Fu TC (2011) A review on time series data mining. Eng Appl Artif Intell 24(1):164–181

Fu TC, Chung FL, Luk R, Ng CM (2004) Financial time series indexing based on low resolution clustering. In: Workshop at the 4th international conference on data mining, pp. 5–14

Fu TC, Chung FL, Luk R, Ng CM (2007) Stock time series pattern matching: template-based vs. rule-based approaches. Eng Appl Artif Intell 20(3):347–364

Fu TC, Chung FL, Luk R, Ng CM (2008) Representing financial time series based on data point importance. Eng Appl Artif Intell 21(2):277–300

Fu TC, Chung FL, Ng CM (2006) Financial time series segmentation based on specialized binary tree representation. In: International conference on data mining, pp. 3-9

Jiang J, Zhang Z, Wang HA (2007) New segmentation algorithm to stock time series based on PIP approach. In: International conference on wireless communications, networking and mobile computing, pp. 5609–5612

Kugiumtzis D, Vlachos I, Papana A, Larsson PG (2007) Assessment of measures of scalar time series analysis in discriminating Preictal states. Int J Bioelectromag 9(3):134–145

Kumiega A, Van Vliet B (2008) Quality money management. Elsevier, Amsterdam

Lo AW, Mamaysky H, Wang J (2000) Foundations of technical analysis: computational algorithms, statistical inference, and empirical implementation. J Financ 55(4):1705–1765

Lucke B (2003) Are technical trading rules profitable? Evidence for head-and-shoulder rules. Appl Econ 35:33–40

Neftci SN (1991) Naïve trading rules in financial markets and Wiener-Kolmogorov prediction theory: a study of "Technical Analysis". J Bus 64(4):549–571

Phetchanchai C, Selamat A, Rehman A, Saba T (2010) Index financial time series based on zigzag-perceptually important points. J Comput Sci 6(12):1389–1395
Savin G, Weller P, Zvingelis J (2007) The predictive power of "Head-and-Shoulders" price patterns in the U.S. stock market. J Financ Econometr 5(2):243–265
Tsinaslanidis PE, Kugiumtzis D (2014) A prediction scheme using perceptually important points and dynamic time warping. Expert Syst Appl 41(15):6848–6860

Chapter 3
Assessing the Predictive Performance of Technical Analysis

3.1 Introduction

This chapter presents some of the celebrated means by which the predictive performance of a technical trading system or a particular technical tool can be assessed. Although not all of these procedures are used in the subsequent chapters, we believe that they are important basic tools for anyone who wishes to assess the performance of such trading systems. It can be argued that TA can be used in two ways. In the first case, individual technical trading rules, or a combination of them, can signal trading recommendations (buy, sell or neutral positions). In the second case, technical tools can be used as explanatory variables in a modeling procedure where the output is returns (or price) predictions (Leigh et al. 2002b; Gencay 1998, 1999). A bundle of methodologies can be used in both cases and are presented in the subsequent sections.

More precisely, the rest of this chapter is organized as follows. In Sect. 3.2 well known methods for assessing the performance of generated trading signals are presented. They include pair tests, Bernoulli trials and bootstrap assessments. Section 3.3 focuses on the means of assessing returns' predictions and presents a bundle of indicators that can be used for measuring either the prediction accuracy and or the predictability of changes in directions. Finally, Sect. 3.4 makes a conclusion.

3.2 Assessing the Performance of Trading Signals

Let $\{P_t\}_{t=1}^{\ell}$ be a price series of length ℓ, where P_t is the closing price on day t and ℓ is the last (most recent) trading day in the examined sample. In addition, assume that our interest is in assessing the performance of a technical trading strategy, $g(\cdot)$ which signals long and/or short trading positions (3.1).

P.E. Tsinaslanidis, A.D. Zapranis, *Technical Analysis for Algorithmic Pattern Recognition*, DOI 10.1007/978-3-319-23636-0_3

$$g(P_t, P_{t-1}, P_{t-2}, \ldots) \equiv g_t = \begin{cases} 1 \; for\ long\ position \\ -1 \; for\ short\ position \\ 0 \; otherwise\ (neutral\ position) \end{cases} \tag{3.1}$$

In this section various means for assessing the performance of such a trading strategy are presented. More precisely, Sect. 3.2.1 discusses the types of holding periods that can be used after the realization of a trading signal. Section 3.2.2 presents the use of pair tests to compare unconditional empirical distributions of returns to the conditional distributions, which are the distributions of returns conditioned on a particular technical trading strategy. Finally, Sect. 3.2.3 describes the use of Bernoulli trials in the context of such assessments, whilst in Sect. 3.2.4 the use of a bootstrap methodology is described.

3.2.1 Defining Holding Periods

The manner the holding periods are defined is crucial in the assessment of the predictive performance of TA. When trading strategies, based on conventions of TA, are erected, holding periods can be defined mainly in two ways.

First, dynamic holding periods can be structured according to the manner the price path evolves after the generated trading signal. For instance, dynamic trading strategies for zigzag patterns presented in Chap. 5, identify three different manners that prices might evolve after a pattern's confirmation. In particular, it is examined whether, prices reach the price target, implied by a technical pattern, within a predefined time limit or evolve differently than TA asserts whereby a stop loss condition is triggered. The event that occurs sooner signals the time the initial position is closed. If none of the above occurs the initial position is closed when the predefined time limit expires. Second, passive (or fixed) holding periods can be adopted instead, where the initial position is closed after a constant predefined time period $h \in \mathbb{N}^+$ (Leigh et al. 2002a, 2004; Wang and Chan 2009; Zapranis and Tsinaslanidis 2012a, b).

3.2.2 Pair Tests

As already mentioned, under the passive holding period scheme, initial trading positions are closed after the predefined constant time period $h \in \mathbb{N}^+ \equiv \{1, 2, \ldots\}$. Algorithmically, we can pre-allocate a vector $\boldsymbol{R^h}$ of size $1 \times \ell$ with *nans* and subsequently define its components, denoted collectively r_t^h, by calculating the generated h-period returns. Equation (3.2) shows r_t^h when arithmetic (or simple) returns are used, $r_t^{h,A}$, whereas (3.3) shows the corresponding equation for logarithmic returns, $r_t^{h,L}$.

$$r_t^{h,A} = \begin{cases} g_t\left(\frac{P_{t+h} - P_t}{P_t}\right) \; if \; g_t \in \{-1, 1\} \\ nan \; if \; g_t = 0 \; or \; t + h > \ell \end{cases} \tag{3.2}$$

$$r_t^{h,L} = \begin{cases} g_t \ln\left(\frac{P_{t+h}}{P_t}\right) \; if \; g_t \in \{-1, 1\} \\ nan \; if \; g_t = 0 \; or \; t + h > \ell \end{cases} \tag{3.3}$$

Subsequently, we keep the indexation regarding the times that trading returns are realised and remove *nans*, reducing the size of $\boldsymbol{R}^{\boldsymbol{h}}$ to $1 \times m$, where $m < \ell$ is the number of trading signals generated by $g(\cdot)$.

Vector $\boldsymbol{R}^{\boldsymbol{h}}$ will be then compared with vector $\boldsymbol{R}^{\boldsymbol{h}^*}$ which contains the h-period arithmetic (3.4) or logarithmic (3.5) returns of the original price series.

$$r_t^{h,A^*} = \begin{cases} \frac{P_{t+h} - P_t}{P_t}, \; for \; t \in \{1, 2, \ldots, \ell - h\} \\ nan \; otherwise \end{cases} \tag{3.4}$$

$$r_t^{h,L^*} = \begin{cases} \ln\left(\frac{P_{t+h}}{P_t}\right), \; for \; t \in \{1, 2, \ldots, \ell - h\} \\ nan \; otherwise \end{cases} \tag{3.5}$$

In particular one-tailed hypothesis pair tests can be conducted for testing the null hypothesis $H_0 : \bar{r}_t^h = \bar{r}_t^{h^*}$, against the alternative $H_1 : \bar{r}_t^h > \bar{r}_t^{h^*}$, where $\bar{r}_t^h$ is the mean return generated by $g(\cdot)$ and $\bar{r}_t^{h^*}$ is the mean market return. Various pair tests can be used for this purpose. It is of prime importance though, that the characteristics of the compared distributions are examined first. If the distributional assumptions of a parametric pair-test are violated, nonparametric pair tests can be alternatively used which are free of the necessity of a particular distributional assumption.

Since we have presented both arithmetic and logarithmic returns in (3.2)–(3.5) it is important to briefly compare the use of these two types of returns in finance applications. First, the difference between $r_t^{h,A}$ and $r_t^{h,L}$ is smaller when small in absolute values returns are considered. In view of finance applications, this implies that, for daily returns whose magnitude is considered generally small, logarithmic returns can be used without introducing great error. Second, logarithmic returns are additive over time which means that having calculated the daily logarithmic returns for the period $[i : T]$, $\{r_t^{1,L}\}_{t=i}^{T}$ (where $i < T$), the logarithmic return corresponding to the whole period is $r_i^{T,L} = \sum_{t=i}^{T} r_t^{1,L}$. Third, if daily logarithmic returns follow a normal distribution, this will also be valid for logarithmic returns of greater periods (weekly, monthly, etc.). The last two aforementioned properties of logarithmic returns are not valid for simple returns. However, in the context of portfolio construction, the arithmetic return of a portfolio is a weighted average of the arithmetic returns of its components which is not the case when logarithmic returns are used. Having this in mind and considering that TA is mainly used for deriving short term trading signals it is usually preferred to use logarithmic returns instead of arithmetic. However, when logarithmic returns are used we advise the reader to examine whether this choice has any impact on his findings.

3.2.3 Bernoulli Trials

In the special case where the trading rule considered is applied on a large number of stocks, Bernoulli trials can be used for assessing its performance. More precisely each stock is treated as a Bernoulli trial with two possible outcomes; a success occurs when the trading rule generates positive profits and a failure where it does not. The number of firms where the trading rule generates positive profits follows a binomial distribution, which converges due to the large number of trials to the normal. If p is the probability of success (a simplified "fair" level of 50 % can be used for this purpose) and N is the number of trials, the mean can be calculated as pN and the sigma as $\sqrt{Np(1-p)}$. Thus, the null hypothesis that the trading rule does not generate positive returns on more stocks that would be expected by chance can be tested by performing a z-test (Marshall et al. 2009). The z-stat can be calculated as $(x - pN)/\sqrt{Np(1-p)}$ where x is the number of successful cases.

3.2.4 The Bootstrap Approach

Bootstrap methodology was initially published by Efron (1979) and it can also be used to assess the performance of a trading rule. In particular, it can be used as an alternative to the statistical inference which is based on the assumption of a parametric model. The study of Brock et al. (1992) is one characteristic example of using bootstrap methodology in assessing the performance of a bundle of technical trading rules. As they have stated, ordinary statistical tests, like the t-ratios they had used, assume normal, stationary and time-independent distributions. However, empirical evidences suggest that financial returns series may exhibit distributional characteristics that deviate from these assumptions, like leptokurtosis, autocorrelation, conditional heteroskedasticity and changing conditional means. This implies that in the latter case, adopting ordinary statistical tests may produce less accurate results than those obtained from a bootstrap methodology, which deals with these issues.

Assume that we are interesting in assessing the performance of a technical trading rule. The bootstrap methodology involves the following procedure. The trading rule considered is applied on the actual price series and the generated profitability is calculated by calculating a statistic like the mean realised return. Subsequently, it is assumed that the actual series is one realization of a null model (like ARIMA or GARCH[1] models) that is used to describe the price generating mechanism. This implies that the actual series is a sample of an unknown population, and the statistic calculated is a sample statistic. Initially, the coefficients of the null model under consideration are estimated. Subsequently, it is examined if the model describes the initial series well, by examining if residuals obtained are independent and identically distributed. These residuals are being resampled with replacement and by using the coefficients of the null model, a large number of simulated series are being generated. These series should have the same distributional characteristics with the initial series but they are random. The technical trading rule is applied on every simulated series and a distribution of the generated mean trading returns is constructed. By comparing the mean return from the actual series with the distribution of the mean returns from the simulated series, simulated p-values are inferred. More precisely, these p-values are fractions indicating the proportion of simulated mean returns that are greater than the mean return realised from the original series. Thus, the trading rule examined would be considered profitable at a significance level of $a\,\%$ if the mean return generated from the actual series is greater than $(100 - a)\%$ of the simulated mean returns (as indicated by a p-value $\leq a$). This methodology is followed and presented in more detail in Chap. 8. Numerous scientific articles have used this approach, like (Marshall et al. 2009; Wang et al. 2014).

3.3 Assessing the Performance of Predicting Returns

Tools of TA can also be used as explanatory variables for predicting returns (or prices). Indicatively, buy and sell signals from simple technical trading rules were used in single layer feedforward networks to predict DJIAI returns in Gencay (1998) and spot foreign exchange rate returns in Gencay (1999). In Leigh et al. (2002b), a feedforward neural network with backpropagation learning was trained with only the trading days occurring at the end of identified (by template matching) bull flags in order to predict the NYSE level. In this section, well known techniques for measuring the predictive performance of such models are being presented.

3.3.1 Measuring the Prediction Accuracy

Let $\{x_j\}_{j=1}^{n+\ell} \equiv \{x_1, x_2, \ldots, x_n, x_{n+1}, \ldots, x_{n+\ell}\}$ be a price series of length $n + \ell$ where the first n observations $\{x_1, x_2, \ldots, x_n\}$ is the initial training set and

[1] Generalised autoregressive conditional heteroskedasticity.

$\{x_{n+1}, x_{n+2} \ldots, x_{n+\ell}\}$ is the initial test set. Additionally, denote $x_t(k)$ the k-steps ahead forecast of the target variable x_{t+k} and $e_t(k) = x_t(k) - x_{t+k}$ the prediction error for $t = \{n, n+1, \ldots, n+l-k\}$. This notation implies that a rolling forecasting procedure is used, whereby on every time t a prediction is made k-steps ahead conditioning all available historical information $\{x_1, x_2, \ldots, x_t\}$.

A celebrated tool for assessing the prediction accuracy is the *Mean Squared Error* (MSE). Equation (3.6) presents the calculation of this measure when predicting k-steps ahead, $MSE(k)$.

$$MSE(k) = \frac{1}{\ell - k + 1}\sum_{j=n}^{n+\ell-k} e_j(k)^2 = \frac{1}{\ell - k + 1}\sum_{j=n}^{n+\ell-k} \left(x_j(k) - x_{j+k}\right)^2 \quad (3.6)$$

In an ideal case, MSE takes its minimum plausible value of 0, implying that the model predicts with perfect accuracy. Of course this ideal case is not plausible in real datasets, but this measure can be used for comparative purposes, when assessing two or more trading rules (or trading systems). In particular, a prediction scheme with lower MSE implies better prediction accuracy than another one with a greater MSE. However, by construction MSE suffers by two main drawbacks. First, it is influenced more heavily by large errors than by small errors. Second, its value is affected by the scale of variable x. This means that, this measure may be inappropriate if used for comparing the predictive accuracy on different variables (e.g. two stocks with prices evolving in different price levels) or different time intervals (e.g. for a particular stock price series, prices may evolve in different levels when different subperiods are examined).

Alternatively, the *Root Mean Squared Error* (RMSE) can be used which is the squared root of MSE (3.7).

$$RMSE(k) = \sqrt{\frac{1}{\ell - k + 1}\sum_{j=n}^{n+\ell-k} e_j(k)^2} = \sqrt{MSE(k)} \quad (3.7)$$

Whilst, MSE illustrates the variance, RMSE corresponds to the standard deviation of the model prediction error. This implies that this measure is expressed in the same units of the response variable, constituting its interpretation an easier task. As in MSE, smaller values indicate better prediction accuracy. However, this measure is still vulnerable to observations' scaling and outliers in errors.

In addition, *Normalized Root Mean Squared Error* (NRMSE) can also be used for measuring the forecasting performance (3.8).

$$NRMSE(k) = \frac{RMSE(k)}{\sqrt{\frac{1}{\ell-k+1}\sum_{j=n}^{n+\ell-k} \left(\bar{x} - x_{j+k}\right)^2}} \quad (3.8)$$

Again, the closer NRMSE is to zero, the better the predictions. The difference in this measure is that the prediction accuracy is compared with a mean-based naïve forecasting model. If $NRMSE \approx 1$, predictions made with the mechanism of consideration are similar to those made by predicting with the mean, $\bar{x}$. Values significantly lower (greater) than 1 imply superiority (inferiority) of the examined forecasting scheme.

Normalized (by) Persistence Root Mean Square Error (NPRMSE) is constructed in the same notion with NRMSE (3.9).

$$NPRMSE(k) = \frac{RMSE(k)}{\sqrt{\frac{1}{\ell-k+1}\sum_{j=n}^{n+\ell-k}\left(x_j - x_{j+k}\right)^2}} \tag{3.9}$$

The difference here is the benchmark of comparison. In particular, a naïve prediction scheme using the last observed value, x_j, for making k-steps ahead predictions is used in the denominator. By construction $NPRMSE(k) \geq 0$ and its interpretation is similar to the one presented for $NRMSE(k)$.

Mean Absolute Error (MAE) measures the prediction accuracy in absolute terms (3.10).

$$MAE(k) = \frac{1}{\ell - k + 1}\sum_{j=n}^{n+\ell-k} \left|e_j(k)\right| \tag{3.10}$$

Compared to MSE and RMSE, this measure of accuracy is less vulnerable to large errors but still depends on the scales. Again, lower values indicate better prediction accuracy.

This last deficiency can be solved with the adoption of *Mean Absolute Percentage Error* (MAPE) where prediction errors are divided by actual values (3.11).

$$MAPE(k) = \frac{100}{\ell - k + 1}\sum_{j=n}^{n+\ell-k}\left|\frac{e_j(k)}{x_{j+k}}\right| = \frac{100}{\ell - k + 1}\sum_{j=n}^{n+\ell-k}\left|\frac{x_j(k)}{x_{j+k}} - 1\right| \tag{3.11}$$

However, this adjustment entails another significant drawback. If $x_{j+k} = 0$, MAPE is undefined. This fact constitutes MAPE inappropriate when predictions are made on price series fluctuating around zero (e.g. daily stock returns).

Finally, another celebrated measure of predictive accuracy is the Theil's U-statistic, U_1, initially proposed in (1961). U_1, for *one*-step ahead predictions, $U_1(1)$, is presented in (3.12).

$$U_1(1) = \frac{\sqrt{\frac{1}{\ell}\sum_{j=n}^{n+\ell-1}\left(x_j(1) - x_{j+1}\right)^2}}{\sqrt{\frac{1}{\ell}\sum_{j=n}^{n+\ell-1}\left(x_j(1)\right)^2} + \sqrt{\frac{1}{\ell}\sum_{j=n}^{n+\ell-1}\left(x_{j+1}\right)^2}} \tag{3.12}$$

When good predictions are made the numerator will have a small value compared to the denominator. In (1966) Theil suggested a second version of U-stat, U_2 (Bliemel

1973; Makridakis et al. 1988; Martínez-Rivera et al. 2012), and (3.13) presents its calculation for *one*-step ahead predictions.

$$U_2(1) = \sqrt{\frac{\sum_{j=n}^{n+\ell-1}\left(\frac{x_j(1)-x_{j+1}}{x_j}\right)^2}{\sum_{j=n}^{n+\ell-1}\left(\frac{x_{j+1}-x_j}{x_j}\right)^2}} \tag{3.13}$$

Theil's U-statistics can be used in two ways; first, as a measure of forecast accuracy (U_1) and second as a measure of forecast quality (U_2). In the first case, $U_1(1)$ takes values in the closed interval $[0, 1]$ with closer values to 0 indicating a better predictive performance. $U_2(1)$ has a similar structure with the *NPRMSE* (1) presented above. Both measures compare the predictive performance of a model with a naïve forecasting method using the most recent observation available as a forecast. The difference is that $U_2(1)$ compares relative prediction errors whereas the *NPRMSE*(1) absolute prediction errors. However, their interpretation is the same. In particular, if $U_2(1) \approx 1$, the examined model predicts similar with the naïve model whilst if $U_2(1)$ is significantly lower (greater) than one, the model outperforms (underperforms) predictions made with the last available price.

3.3.2 Measuring the Predictability of Changes in Directions

This section presents a bundle of indicators which measure the ability of the examined model to predict changes regardless their size. These measures are the *Prediction of Change in Direction* (POCID), the *Independent Prediction of Change in Direction* (IPOCID) and the *Prediction of Sign* (POS). They are usually expressed as percentages and values towards 100 % indicate 'good' predictive performance.

All these three measures are calculated using (3.14).

$$\{POCID(k), IPOCID(k), POS(k)\} = \frac{100}{\ell - k + 1}\sum_{j=n}^{n+\ell-k} d_j \tag{3.14}$$

What changes in the calculation of every measure is the binary variable, $d_j \in \{0, 1\}$, used in the summation. In particular, for the calculation of *POCID*(k), d_j is given by (3.15).

$$d_j = \begin{cases} 1, & \textit{if } (x_{j+k} - x_j)(x_j(k) - x_j) > 0 \\ 0, & \textit{if } (x_{j+k} - x_j)(x_j(k) - x_j) \le 0 \end{cases} \tag{3.15}$$

Similarly, Eqs. (3.16) and (3.17) define d_j for the *IPOCID*(k) and *POS*(k) respectively.

$$d_j = \begin{cases} 1, \text{ if } (x_{j+k} - x_j)(x_j(k) - x_{j-k}(k)) > 0 \\ 0, \text{ if } (x_{j+k} - x_j)(x_j(k) - x_{j-k}(k)) \leq 0 \end{cases} \tag{3.16}$$

$$d_j = \begin{cases} 1, \text{ if } x_{j+k}x_j(k) > 0 \\ 0, \text{ if } x_{j+k}x_j(k) \leq 0 \end{cases} \tag{3.17}$$

POCID and IPOCID are generally used in applications where predictions are made on a price path. Contrary, POS is mainly adopted in cases where the model's prediction focuses on returns (i.e. the sequence $\{x_j\}$ is a series of relative changes).

3.3.3 Scatter Plots and Linear Regression Between Targets and Predictions

Another way to assess the k-step ahead predictive performance of a model is to apply a linear regression (3.18) between predictions, $\{\hat{y}_t\}_{t=k}^{\ell} \equiv \{x_j(k)\}_{j=n}^{n+\ell-k}$ and target variables (actual observations), $\{y_t\}_{t=k}^{\ell} \equiv \{x_j\}_{j=n+k}^{n+\ell}$.

$$y_t = b_0 + b_1 \hat{y}_t + e_t \tag{3.18}$$

The presence of linear relationship between actual values and predictions can be examined by testing the null hypothesis that b_1 is zero, $H_0 : b_1 = 0$, against the alternative $H_1 : b_1 \neq 0$. Assuming normal distribution for the residuals, e_t, t-statistic will follow the t-Student distribution with $v = \ell - k - 1$ degrees of freedom and will be defined as

$$t_{(v)} = \frac{b_1}{s(b_1)}. \tag{3.19}$$

In (3.19), $s(b_1)$ is the standard error of coefficient b_1 given by (3.20).

$$s(b_1) = \sqrt{\frac{\frac{1}{\ell-k-1}\sum_{t=k}^{\ell} e_t^2}{\sum_{t=k}^{\ell} (\hat{y}_t - \bar{\hat{y}})}} \tag{3.20}$$

Here the numerator is the standard error of the residuals and $\bar{\hat{y}}$ is the mean prediction. However, although this hypothesis test is useful in identifying whether there is a linear relationship between actual values and predictions it is also required to test whether the line in describing the above relationship passes from the axes intersection and whether it has an angle of 45°. In the first case the null hypothesis that the constant b_0 is zero is tested against the alternative that b_0 is different than zero. In the second case the hypothesis tested are $H_0 : b_1 = 1$ against $H_1 : b_1 \neq 1$.

Other measures from the simple linear regression analysis, like the coefficient of determination, R^2, and the analysis of variance (ANOVA), like F-stat, can be used to evaluate the linear relationship of two variables. In an ideal (unrealistic) scenario whereby the model perfectly predicts actual values, $\hat{y}_t = y_t, \forall t, b_0 = 0, b_1 = 1$ and $R^2 = 1$.

3.4 Conclusions

In this chapter a bundle of well-known methods for assessing the performance of a technical trading system are presented. The purpose of this chapter was to discuss on a selection of techniques, found in the literature, that can be used in the assessment procedure. Although not all of them are used in the later chapters of this book, we believe that they can serve as a starting point for someone interested in this field.

References

Bliemel F (1973) Theil's forecast accuracy coefficient: a clarification. J Mark Res 10(4):444–446
Brock W, Lakonishok J, Lebaron B (1992) Simple technical trading rules and the stochastic properties of stock returns. J Financ 48(5):1731–1764
Efron B (1979) Bootstrap methods: another look at the Jackknife. Ann Stat 7(1):1–26
Gencay R (1998) The predictability of security returns with simple technical trading rules. J Empir Financ 5(4):347–359
Gencay R (1999) Linear, non-linear and essential foreign exchange rate prediction with simple technical trading rules. J Int Econ 47(1):91–107
Leigh W, Paz N, Purvis R (2002a) Market timing: a test of a charting heuristic. Econ Lett 77:55–63
Leigh W, Purvis R, Ragusa JM (2002b) Forecasting the NYSE composite index with technical analysis, pattern recognizer, neural network, and genetic algorithm: a case study in romantic decision support. Decis Support Syst 32:361–377
Leigh W, Modani N, Hightower R (2004) A computational implementation of stock charting: abrupt volume increase as signal for movement in New York Stock Exchange Composite Index. Decis Support Syst 37:515–530
Makridakis S, Wheelwright SC, Hyndman RJ (1988) Forecasting methods and applications. Wiley, New York
Marshall BR, Qian S, Young M (2009) Is technical analysis profitable on US stocks with certain size, liquidity or industry characteristics? Appl Financ Econ 19:1213–1221
Martínez-Rivera B, Ventosa-Santaulària D, Eduardo Vera-Valdés J (2012) Spurious forecasts? J Forecast 31(3):245–259
Theil H (1961) Economic forecasts and policy. North-Holland, Amsterdam
Theil H (1966) Applied economic forecasting. North-Holland, Amsterdam
Wang J, Chan S (2009) Trading rule discovery in the US stock market: an empirical study. Expert Syst Appl 36:5450–5455
Wang F, Philip LH, Cheung DW (2014) Combining technical trading rules using particle swarm optimization. Expert Syst Appl 41(6):3016–3026

Zapranis A, Tsinaslanidis PE (2012a) Identifying and evaluating horizontal support and resistance levels: an empirical study on US stock markets. Appl Financ Econ 22:1571–1585. doi:10.1080/09603107.2012.663469

Zapranis A, Tsinaslanidis PE (2012b) A novel, rule-based technical pattern identification mechanism: Identifying and evaluating saucers and resistant levels in the US stock market. Expert Syst Appl 39(7):6301–6308. doi:10.1016/j.eswa.2011.11.079

Chapter 4
Horizontal Patterns

4.1 Introduction

Support and resistance levels (SARs) are two celebrated tools of TA. A support level is defined always under the spot price of a financial asset. It is rational to set such a level when the price evolves in a bearish[1] manner. It is the price level that the price finds a support when it reaches it from above. When this happens, a reverse in the trend is expected, especially when this reversal is accompanied by a large increase in the volume of transactions. In other words investors think that the corresponding asset is undervalued at this level so the demand is being instantly increased and buyers support the corresponding stock at that level. The reverse stands for a resistance level. Contrary, if price penetrates downwards (upwards) a support (resistance) level the initial trend is expected to continue. In the first case support and resistance levels act as trend reversal predictors where in the second case as trend continuation predictors.

They are widely used by practitioners either isolated or as components of a variety of chart patterns. For example, the neckline of the "Head and Shoulders" pattern is considered as support level before its penetration and as a resistance afterwards (see Chap. 5). Additionally, they can be signified by technical indicators (e.g. a long term moving average). In most studies, SARs are identified as minima or maxima realised in a precedent, constant time interval (Brock et al. 1992; Hudson et al. 1996; Raj and Thurston 1996; Curcio et al. 1997; Coutts and Cheung 2000; Qi and Wu 2006; Marshall et al. 2008a, b; 2009; De Zwart et al. 2009; Mizrach and Weerts 2009; Cheung et al. 2011; Falbo and Cristian 2011). Osler

[1] In technical jargon a bull trend is associated with augmented investors' confidence combined with the analogous increase in the investing, due to the expectation of future price increases. A bear trend is the reverse. The use of "bull" and "bear" to describe market trends, stems from the manner that those two animals attack their opponents. Thus an uptrend is characterised as bullish and a downtrend as bearish.

P.E. Tsinaslanidis, A.D. Zapranis, *Technical Analysis for Algorithmic Pattern Recognition*, DOI 10.1007/978-3-319-23636-0_4

(2000) mentions that this is the simplest approach to identify SAR visually and this conforms to Murphy (1986) who states that "*usually, a support level is identified beforehand by a previous reaction low*". Kavajecz and Odders-White (2004) define support levels as the lowest bid prices observed during the last week, given that these locals are realized at least twice during that period.

In their purest form, SARs can be defined as specific price levels. In this case we can call them *horizontal support and resistance levels* (HSARs) due to the practice to draw them as horizontal lines on the chart. Several variations of HSARs exist which can be classified by the manner a technical analyst identifies them. Osler (2000) reports that these ways are, simple numerical rules, inside information, market psychology, and visual assessment. In this chapter, emphasis is given in the latter one because, as mentioned before, one of the aims of this book is the proposition and implementation of novel, rule-based identification mechanisms for a variety of technical patterns, that enclose the principles found in the identification process via visual assessment.

The rest of the chapter is organized as follows. In Sect. 4.2, four means of identifying HSARs are presented. In Sect. 4.3, our proposed algorithmic technique for identifying HSARs is presented (Zapranis and Tsinaslanidis 2012a). The manner, by which the predictive performance of this technique is assessed, is presented in Sect. 4.4. In Sect. 4.5, empirical results obtained from this analysis are being reported. Finally, in Sect. 4.6 we conclude. In Appendices 1 and 2 the reader will find the Matlab code we developed for the methodology presented in Sect. 4.3.

4.2 Existed HSARs Identification Techniques

As already mentioned, various ways of determining HSARs can be found in technical manuals. This section presents briefly four ways other than the one we scrutiny further in Sect. 4.3. More precisely, the identification of HSARs by using simple numerical rules is presented in Sect. 4.2.1. In Sect. 4.2.2 the use of public announcements or inside information for deriving these levels is discussed. The effect of market psychology in determining HSARs is discussed in Sect. 4.2.3. Finally, in Sect. 4.2.4 a celebrated version of HSARs, known in TA as trading range breakouts, is presented.

4.2.1 HSARs Identified by Simple Numerical Rules

In this case a technical analyst identifies initially an uptrend (or downtrend). HSARs are drawn in a manner that they pass through specific levels between the two extreme points that characterize this trend. For example, Osler (2000) refers to the "50 percent rule" which affirms that after a significant price move, the direction at which a price series evolves, is going to be inverted at about 50 % of the original move. Similarly is the process when a technical analyst applies the "one-third" and "two-thirds" rule.

The Fibonacci number sequence is $\{F_n\} = \{0, 1, 1, 2, 3, 5, 8, 13, 21, \ldots\}$ where every number results by adding the two preceding numbers. Equation (4.1) presents this recurrence relation, where $n \in \mathbb{N}_0$, and the two initial seed values are $F_0 = 0$ and $F_1 = 1$. The analytic relation is presented in (4.2) while the analytic relation with the use of the golden ratio $\varphi = \left(1 + \sqrt{5}\right)/2 = 1.61803\ldots$ is given in (4.3).This sequence of numbers is being characterized by many other known and interesting relationships. Murphy (1986) refers that some of these relationships were known to the Greeks, the Egyptians and Leonardo da Vinci. It is generally accepted though that these relationships are embedded and reappear in many nature's phenomena and human's activities. Since, one of the human's activities is the investment activity, Fibonacci numbers has been used to deduce psychological support and resistance price levels. These levels are 0.0 %, 23.6 %, 38.2 %, 50 %, 61.8 %, 100 %, 161.8 %, 261.8 %, and 423.6 % and are known as Fibonacci retracements (Achelis 1995).[2]

$$F_n = F_{n-1} + F_{n-2}, \quad \forall n \in \mathbb{N}_{>1} \tag{4.1}$$

$$F_n = \frac{1}{\sqrt{5}}\left(\frac{1+\sqrt{5}}{2}\right)^n - \frac{1}{\sqrt{5}}\left(\frac{1-\sqrt{5}}{2}\right)^n, \quad \forall n \in \mathbb{N}_0 \tag{4.2}$$

$$F_n = \frac{\varphi^n - (1-\varphi)^n}{\sqrt{5}}, \quad \forall n \in \mathbb{N}_0 \tag{4.3}$$

Figure 4.1 illustrates Fibonacci retracements drawn on the IBM stock price series. Based on the significant uptrend which occurred at the period 08/05/2003–02/04/2004 Fibonacci retracements were drawn according to the aforementioned ratios. At the subsequent subperiod (02/04/2004–02/28/2005) prices retrace, finding support and resistance at or near Fibonacci retracement levels.

4.2.2 HSARs Identified with Public Announcements or Inside Information

Information used in this case concerns mainly large trading orders set by wealth investors. For example, if someone is informed that for tomorrow, a large long position is set by a firm on a specific stock's price level, lower than today's level, then he would reasonably treat the order's price as a support level. Such information is mainly private or circulated informally in the market and is available for market

---

[2] The ratios 61.8 %, 38.2 % and 23.6 % are approximated if we divide F_n by F_{n+1}, F_{n+2} and F_{n+3} respectively. The 50 % level is not an actual Fibonacci level. However, it is originates from the Dow Theory's assertion that the prices often retrace half their prior move.

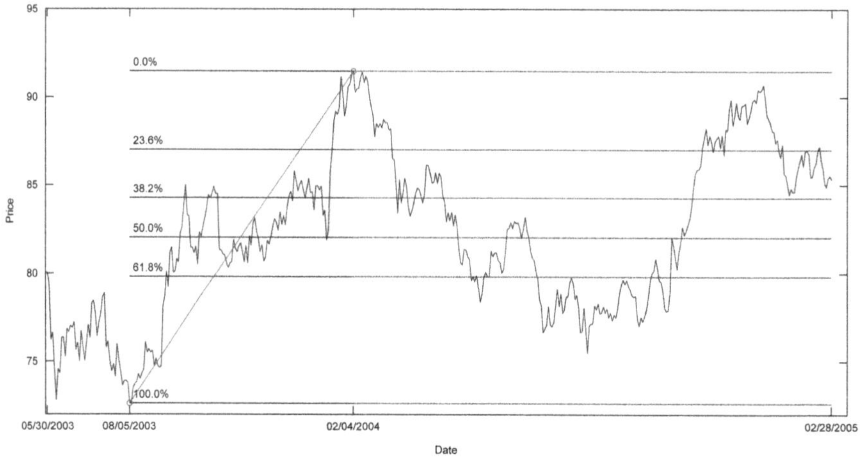

Fig. 4.1 Fibonacci retracements drawn on IBM stock price series (period from 05/30/2003 to 02/28/2005)

insiders (Osler 2000). However it is plausible that this information can be found in publicly available announcements. In both cases, the measurement of the predictive power of such support and resistance levels can be considered as tests for the semi-strong and strong form efficiency of a market, and not for a weak form, since information used here is not deduced from historical prices.

4.2.3 HSARs Based on Market Psychology

The creation of support and resistance levels may also be attributed to psychological reasons. It is observed in various studies that, a trader has an innate tendency to think in terms of round prices, each time he is about to take a long or short position. This behavior is also observed by Goodhart and Figliuoli (1991) who examined bid-ask spreads for major currencies and they refer to it as a "*round number syndrome*". In case these round figures constitute historical long-term peaks or bottoms, the validity of the generated SARs increases (Edwards and Magee 1997). This round number syndrome is also known in the academic literature as psychological barriers (De Grauwe and Decupere 1992; Dorfleitner and Klein 2009; Mitchell 2001, 2006).

4.2.4 Trading Range Breakouts

Another celebrated version of SARs is the trading range breakout (TRB). The efficacy of TRB has been evaluated in many previous studies (Brock et al. 1992;

Curcio et al. 1997; Zapranis and Tsinaslanidis 2012b). When TRB used as a trading strategy, a support (resistance) level is set by the minimum (maximum) closing price realized in a previous time period "w". This strategy can be extended by incorporating various filters such as a percentage-band filter, a time delay filter, a stop-loss position and volume confirmation (Qi and Wu 2006). These filters are used as tools of confirmation. In other words, a trader confirms the validity of the penetration of a support (resistance) level when one or more of the aforementioned filters are satisfied. The above description implies that this trading strategy can be easily quantified by (4.4) and (4.5).[3]

$$Support_t = min\{p_{t-1}, p_{t-2}, \cdots p_{t-w}\} \tag{4.4}$$

$$Resistance_t = max\{p_{t-1}, p_{t-2}, \cdots p_{t-w}\} \tag{4.5}$$

Here the support (resistance) level at time t equals with the minimum (maximum) closing price, p_{t-n} (where $n < w$) observed in the preceding time interval of length w. In case these levels are penetrated, the corresponding investment recommendations are given by the following sets of criteria (4.6) (De Zwart et al. 2009).

$$Investment\ Recomendation \begin{cases} Buy\ if\ \ p_t > Resistance_t \\ Sell\ if\ \ p_t < Support_t \\ Neutral\ otherwise \end{cases} \tag{4.6}$$

4.3 Identifying Horizontal Support and Resistance Levels (HSARs)

The purpose of this section is to propose a novel identification algorithm that captures and quantifies important descriptive conventions of TA which define the identification of HSARs via visual assessment. For instance, Murphy (1986) reports:

> …support is a level or area on the chart under the market where buying interest is sufficiently strong to overcome selling pressure. As a result, a decline is halted and prices turn back up again…Resistance is the opposite of support.

This description is unambiguously correct but it could refer only to regional locals. Bulkowski (2002) completes:

> Perhaps a better definition would include several minor lows or minor highs stopping near the same price multiple times.

By the phrase "near the same price" the definition becomes more realistic since in practice it is extremely rare to observe several locals at the exact price level.

[3] The estimation of this type of SARs is feasible with the following two Matlab functions included in the financial toolbox: $hhigh(\cdot)$ and $llow(\cdot)$. These functions return a vector of highest highs and lowest lows respectively observed the past w periods. The input variable w is specified by the user.

Although technical analysts follow the usual practice to report HSARs as individual price levels, we would have to agree with Bulkowski (2002) who provides a more comprehensive description. Particularly, he observes:

> Support and resistance are not individual price points, but rather thick bands of molasses that slow or even stop price movement.

Finally another characteristic of HSARs is that after breaching a support (resistance) level, this level becomes a resistance (support) level. Achelis (1995) mentions:

> Following the penetration of a support/resistance level, it is common for traders to question the new price levels.

In this case, prices return to the preceding HSAR, but its role switches from support (resistance) to resistance (support). Achelis attributes this behavior in a phenomena he calls "trader's remorse".

Although the simplest way to identify HSARs at a given time is by seeking within a constant, preceding time period for one local minima or maxima (see Sect. 4.2.4), the subsequent proposed methodology (Zapranis and Tsinaslanidis 2012a) captures the aforementioned statements which infer that a HSAR is an area of prices in where local peaks and bottoms reside, rather than a specific individual price level.

The core concept of this identification process includes the classification of all regional locals identified by the $RW(\cdot)$ function (see Chap. 2) into discrete price intervals (bins) and the calculation of the corresponding frequencies. Each bin would represent an area of prices and when its frequency is greater than two, this bin would signify a *horizontal support and resistance zone* (HSARz). The same practice is used to create an ordinary histogram with bins of equal width. The drawback of this method is that the width of bins represents price intervals, which means that the percentage distance of these bins would be different from cluster to cluster. For example, when observing the frequencies of peaks and bottoms between two bins (b_a and b_b), whose bounds are 10–11 and 20–21 respectively, the corresponding percentages would be 10 % and 5 % respectively. In other words, in the first case a HSARz is defined which includes locals that can distance greater in percentage terms, than those included in the second HSARz. This fact increases the probability to observe regional locals in HSARz set at lower price levels. To overcome this problem, the $HSAR(\cdot)$ function is erected (see Appendix 1).

$$[SAR, Bounds, Freq, x\_act] = HSAR(ys, w, x, pflag)$$

The $HSAR(\cdot)$ function takes four input variables: *ys*, *w*, *x* and *pflag*. Since $RW(\cdot)$ is embedded in this function *ys* and *w* are defined as in $RW(\cdot)$. The third input, *x*, is the desired percentage distance of each bin.[4] Finally, if *pflag* equals to unity the

[4] Although, theoretically $x \in \mathbb{R}_{>0}$ in practice we should set this ratio to a logical bound say $(0, 1]$ depending on the examined price series.

corresponding graphs are returned. The algorithmic process is described as follows. Prices of k regional locals (both peaks and bottoms) identified by $RW(\cdot)$ construct a $k \times 1$ column vector $\boldsymbol{L}$ (4.7).

$$\boldsymbol{L} = [l_1, l_2, l_3 \ldots, l_k]' \tag{4.7}$$

Denote L_1^* and L_2^* as:

$$L_1^* = \frac{\min(\boldsymbol{L})}{1 + \frac{x}{2}} \tag{4.8}$$

and

$$L_2^* = \max(\boldsymbol{L})\left(1 + \frac{x}{2}\right). \tag{4.9}$$

Equations (4.8) and (4.9) center the minimum and maximum identified regional local to the first and last bin we are about to create. The number of bins, n, should satisfy (4.10).

$$L_1^*(1 + x)^n = L_2^* \tag{4.10}$$

Solving by n gives

$$n = \log_{(1+x)}\left(\frac{L_2^*}{L_1^*}\right). \tag{4.11}$$

Considering a basic logarithmic property (4.12),

$$\log_a b = \frac{\log_c b}{\log_c a} \tag{4.12}$$

(4.11) becomes

$$n = \frac{\ln\left(\frac{L_2^*}{L_1^*}\right)}{\ln(1 + x)}. \tag{4.13}$$

The actual number of bins $(\hat{n})$ must be an integer number and it results by rounding n to the closest integer. Due to this change variable x has to be reconsidered. The new estimated percentage distance of each bin, $\hat{x}$, must satisfy (4.14).

$$L_1^*(1 + \hat{x})^{\hat{n}} = L_2^* \Rightarrow \hat{x} = \sqrt[\hat{n}]{\frac{L_2^*}{L_1^*}} - 1 \tag{4.14}$$

Closing prices, adjusted for dividends and splits, of Microsoft Corporation (MSFT) traded at NASDAQ for the period 3-Jan-2000 to 1-Dec-2010 are illustrated in Fig. 4.2. When a rolling window of 60 days and a desired percentage bin distance (x) of 5 % is applied $HSAR(\cdot)$ identifies 9 HSARs. The corresponding histogram is presented in Fig. 4.3 showing the 16 discrete price intervals ($\hat{n}$) of different width. These differences in bins' widths results because we keep the actual percentage distance ($\hat{x}$) at a constant level which is 5.12 %. Only price areas (bins) with frequencies greater or equal to two are considered as HSAR. The bins' mean prices which correspond to the identified HSAR are tabulated in Table 4.1.

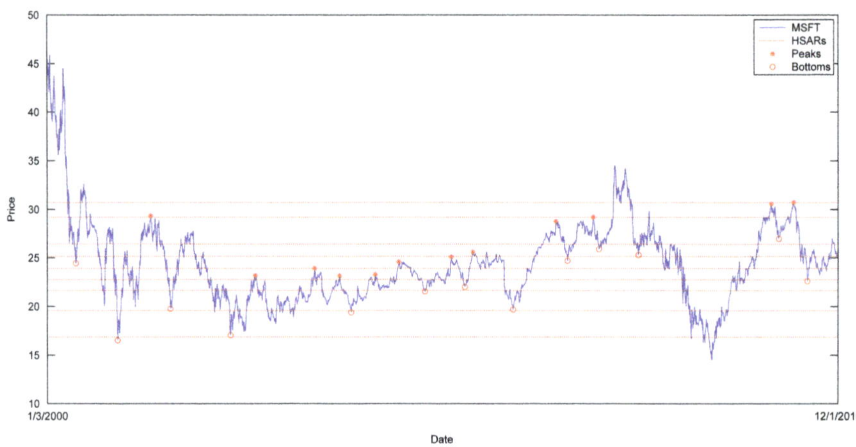

Fig. 4.2 Implementation of $HSAR(\cdot)$ on MSFT ($w = 60,\ x = 5\ \%,\ \hat{n} = 16,\ \hat{x} = 5.12\ \%$)

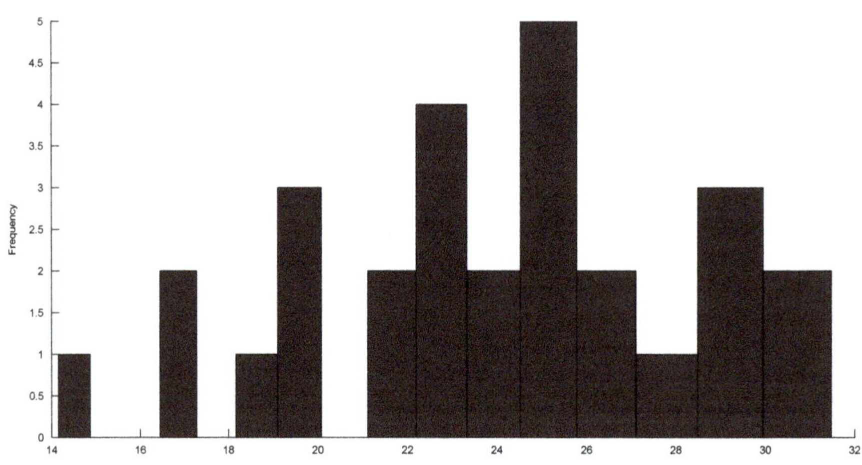

Fig. 4.3 Histogram presenting the number of regional locals used for the identified HSARs presented in Fig. 4.2

Table 4.1 Descriptive data for Figs. 4.2 and 4.3

$HSAR_{id}$	1	2	3	4	5	6	7	8	9
HSAR	16.87	19.59	21.65	22.76	23.92	25.15	26.44	29.21	30.71
Frequency	2	3	2	4	2	5	2	3	2

HSAR horizontal support and resistance level, $HSAR_{id}$ index for HSARs identified

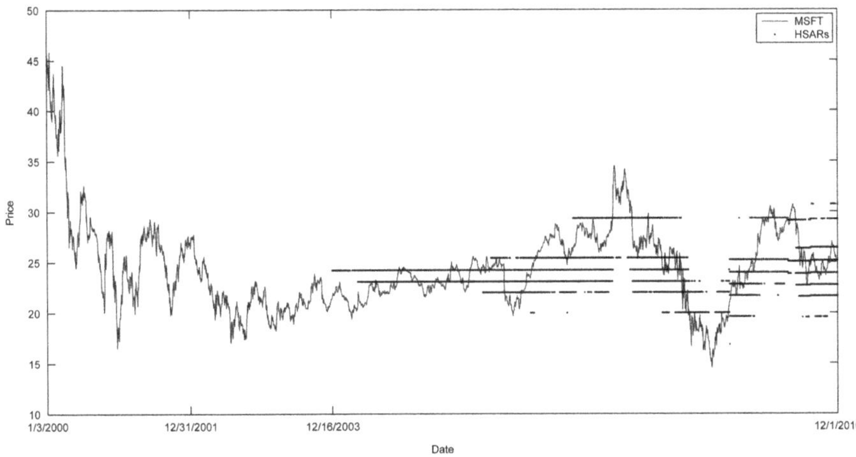

Fig. 4.4 HSARs identified on MSFT on a daily basis. To simplify the illustration only HSARs distanced within ±20 % of each daily price are presented

Until this point, the aforementioned process lacks validity since it uses future information to derive HSARs. Any effort to evaluate the predictive power of these HSARs is irrational. More concretely, we cannot evaluate the performance of a given HSAR, estimated at time *t*, if locals observed after this time are used for its identification. Regarding the MSFT example identified HSARs presented in Table 4.1 should be used for the period after 12/1/2010. To deal with this problem, the daily HSARs are drawn with the following developed function (see Appendix 2):

$$[Results] = HSARsim(ys, w, x, indata, pflag)$$

Firstly a preliminary process is necessary where, for each stock price series the first 2 years (approximately 500 trading days) are used to identify existed horizontal HSARs with the $HSAR(\cdot)$ function. Subsequently, on every successive day (t) the same function estimates the new HSARs using all closing prices from the initial till the last available observation $(t-1)$. Figure 4.4 illustrates the HSARs estimated on a daily basis for MSFT security. By setting $w = 60$ and $x = 5\%$, the algorithm identifies HSARs firstly at the 994th observation (and not at the 501st). This is attributed to the fact that regional locals identified at the first 500 days (from 1/3/2000 to 12/31/2001) failed to form a cluster with frequency greater than 2. In order to accomplish this, additional 493 days were needed, with the algorithm returning the first HSAR on 12/16/2003.

4.4 Assessing the Predictive Performance

The proposed rule-based identification mechanism (Sect. 4.3) is applied on the stocks that compose the NASAQ and NYSE indices. On August 2010 a dataset of 3803 stocks from NASDAQ and NYSE was composed. The data used are daily adjusted closing prices, and the requested time period for the dataset's construction is from July 31, 1990 to July 31, 2010. A filter similar to that implemented in Lo et al. (2000), Marshall et al. (2009) and Mizrach and Weerts (2009) was used. The stocks that had no available closing prices at all, and the stocks that had no available closing prices for the years requested were excluded. Furthermore, for each individual stock, the number of nontrades for each year was counted. Stocks that presented at least 1 year with missing values at a percentage greater than 5 % were also excluded. After implementing these filters, the remainder dataset consists of 232 stocks from NASDAQ and 501 stocks from NYSE. Remaining missing values were completed with linear interpolation applied on the available values realized before and after the missing ones.

The evaluation of the HSARs' performance presented in this chapter, consists of two tiers. The first one examines if these levels can successfully predict trend reversals in a manner similar to the one presented in Osler (2000). The second tier examines if trading based on the signals that these levels generate can be characterized as profitable. All the tests were implemented on the whole sample period and on three, equally length, nonoverlapping subperiods of 6 years,[5] to reduce data snooping biases (Brock et al. 1992; Chong and Ng 2008).

Assume that on day t the price p_t is found to be within the bounds of a support or resistance level. If the previous day's closing price, P_{t-1}, was above (below) the HSARz then a *hit* occurs and a case of support (resistance) level is realized. It is also possible for P_{t-1} to be inside this zone. In such case the comparison is made with the nearest previous observation (P_{t-k}) that was outside this zone. Subsequently, the first forthcoming day $(t+m)$ where price breaches the HSARz either upwards or downwards is considered. In case of a support level when this penetration is upwards (downwards) then a *bounce* (*failure*) occurs. The converse stands for a resistance level. We kept the terminology used in Osler (2000) and the ratio of times the price bounces to the number of actual hits will be referred hereafter as *bounce frequency*. Obviously, a bounce signifies a trend interruption and a failure represents a trend continuation. A theoretical example of this process, regarding a

[5] The first 2 years are used for the preliminary estimation of HSARs for the first sub-period and for the whole sample period. The second and third sub-periods of comparison use for the same purpose all previous, available information.

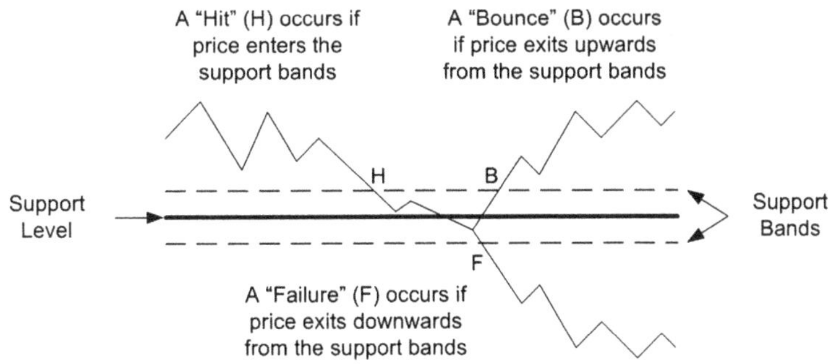

Fig. 4.5 Theoretical example of a horizontal support zone

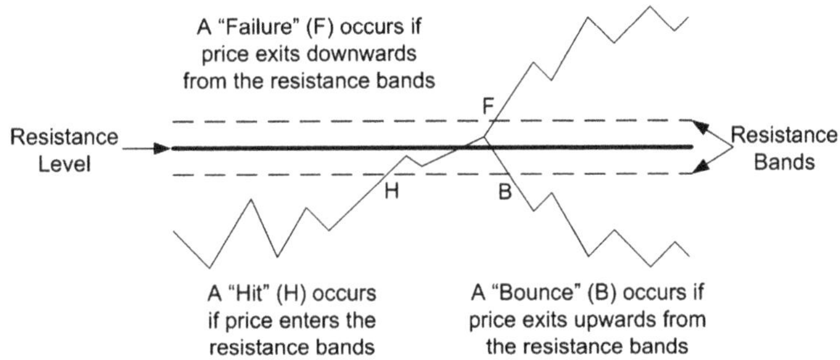

Fig. 4.6 Theoretical example of a horizontal resistance zone

support case is illustrated in Fig. 4.5. The corresponding illustration for a resistance level is presented in Fig. 4.6.

The evaluation of estimated HSARs, concerning their trend-interruption predictability, is implemented by comparing the generated bounce frequencies with those realized by artificially generated ones. More precisely, on every trading day, ten randomly generated HSARs (5 for supports and 5 for resistances) are implemented using (4.15).

$$HSAR_{t,i} = p_t(1 \pm z_{t,i} \times range) \tag{4.15}$$

Here, t represents the trading day, $HSAR_{t,i}$ is the ith artificial HSAR at day t, $i \in [1 : 10]$, p_t is the closing price and $z_{t,i}$ are pseudorandom numbers, drawn randomly with replacement from a uniform distribution, i.e., $z_{t,i} \sim U(0, 1)$. The *range* for a given stock is the largest absolute daily return observed in that stock price series within the subperiod of comparison. By this process, for a given price series, approximately 45,000 artificial supports and resistance levels are generated.[6]

[6] The number 45,000 results by creating 10 artificial HSARs, for every one of the 250 trading days per year, for 18 years.

Even if HSARs prove to be able to predict trend reversals, what is of prime importance is to examine if trading signals generated by these levels are able to realize systematically, statistical significant excess returns. For this analysis two different trading rules are implemented. In the first one (TR_1) a trade signal occurs after a bounce with one day lag to avoid cases of nonsynchronous trading (Bessembinder and Chan 1995; Wang and Chan (2007, 2009). The type of the initial position depends on the type of the HSAR. More precisely, in case price bounces on a support (resistance) level an initial long (short sell) position is taken respectively. This trading rule refers to those investors who believe that there are more chances for a support or resistance level to act as a trend-reversal, rather than a trend-continuation indicator. Conversely, for investors who prefer investing after HSARs are being penetrated, we use a second trading rule (TR_2) which signals a long (short sell) position after price breaches a resistance (support) level (case of failure in Figs. 4.5 and 4.6) (De Zwart et al. 2009). Likewise, in this trading rule the initial positions are also set with one day lag. In both trading rules the initial positions are closed after fixed sets of holding periods (HP). Two sets are used for this purpose; one short term (HP_s) and one medium term (HP_m).

$$HP_s = \{1 : 1 : 20\} \tag{4.16}$$

$$HP_m = \{22 : 2 : 40\} \tag{4.17}$$

HP_s set includes values of 1 until 20 with a step of 1. This means that initial positions are closed after 1, 2, 3, . . ., 20 days. The second set of holding periods can be interpreted analogously. Therefore, a total of 30 different holding periods are used at each trading rule. In cases where the day that the initial position has to be closed is beyond the available sample period, the position is closed on the last available observation. This rare situation can occur in case that a trade signal is triggered at the last available observations of the dataset used. Returns generated by the above trading rules are compared with the corresponding market returns in a manner similar with the one used in Leigh et al. (2004) and Wang and Chan (2007, 2009). Empirical results of this analysis are presented in the Sect. 4.5.

4.5 Empirical Results

This section evaluates firstly the ability of HSARs to predict trend reversals (an analysis we call "bounce analysis", Sect. 4.5.1), and secondly examines whether trading signals are able to generate systematically abnormal returns ("profitability analysis", Sect. 4.5.2). Results presented in this section correspond to HSARs identified with a rolling window of 101 days ($w = 50$) and a desired bin's distance, x, of 3 %.

4.5.1 Bounce Analysis

The first part of this analysis includes the evaluation of HSARs as trend-reversal predictors. Osler (2000) reports that published levels outperform artificial ones by comparing the relative bounce frequencies. A similar approach is followed in this section and results confirm the aforementioned finding in both NASDAQ and NYSE exchanges. By scrutinizing further these results it is found that, in both exchanges, only support levels actually predict trend reversals statistical significantly. In fact the superior performance of support levels compensates for the poorer performance of resistance levels, and this constitutes the overall picture of HSAR as efficient (Figs. 4.7 and 4.8). Inspired by the title of Osler (2000), based on our findings, this part of this analysis could be titled as "supporting supports and resisting resistances".

For each stock the corresponding bounce frequency is calculated for the estimated, B^e, and artificially, B^a, created HSARs. Subsequently, we count the number of stocks where the estimated HSARs realize greater bounce frequencies than those created from the artificial ones. Purporting that estimated HSARs are not efficient, this proportion should not be substantially greater than a probability of 50 % which is attributed to pure chance. In such case the probability of observing $B^e > B^a$ will conform to the binomial distribution.

More concretely we can view each stock as a "Bernoulli trial" in which $B^e > B^a$ is a random event with a probability of success of 50 %. For both exchanges and for all subperiods of examination, the bounce frequencies of the combined estimated support and resistance levels and of the isolated estimated support levels

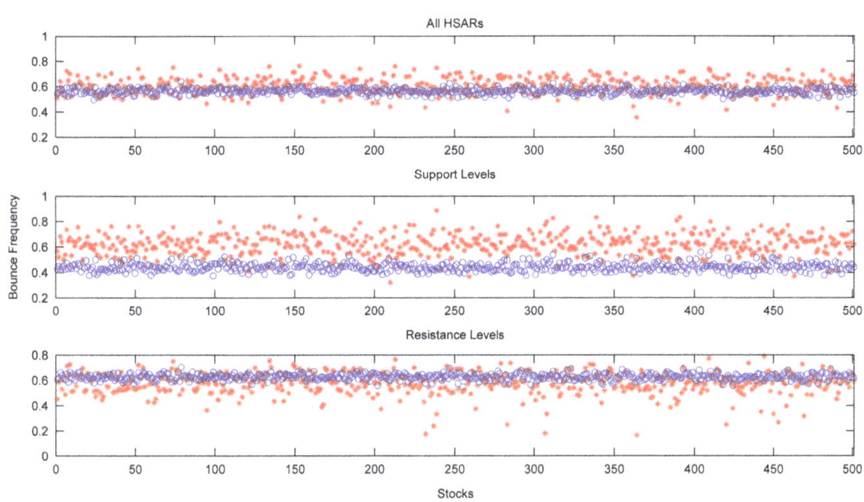

Fig. 4.7 Bounce frequencies for NYSE components. *Stars* symbolize bounce frequencies created by estimated HSARs, and *circles* those created by artificial ones. The first sub-plot corresponds to the combined picture of all HSARs, where the other two illustrate bounce frequencies for support and resistance levels separately

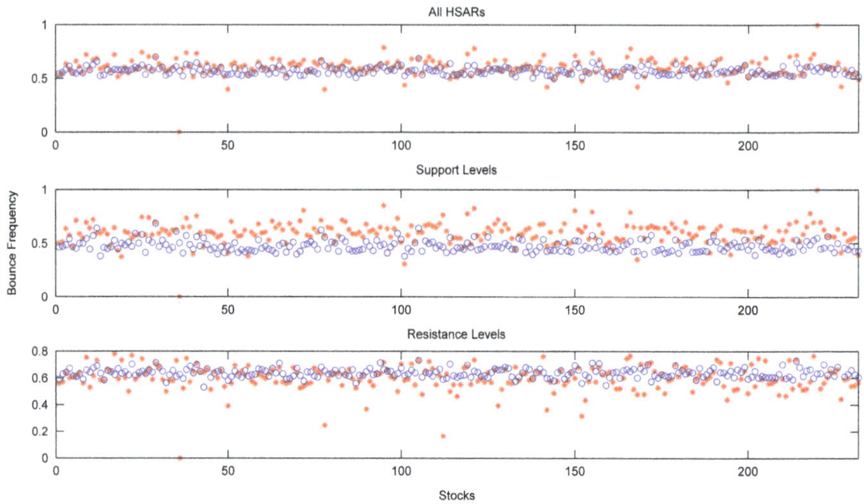

Fig. 4.8 Bounce frequencies for NASDAQ components (same description with Fig. 4.7)

Table 4.2 Bounce analysis results for HSARs

	$B^e > B^a$	Mean	Sigma	z-Stat.	p-Value
Panel A: NYSE					
All HSARs	384/501	250.5	11.19	11.929	0.000
Supports	484/501	250.5	11.19	20.864	0.000
Resistances	121/501	250.5	11.19	−115.713	1.000
Panel B: NASDAQ					
All HSARs	176/232	116	7.62	7.878	0.000
Supports	213/232	116	7.62	12.737	0.000
Resistances	72/232	116	7.62	−57.775	1.000

$B^e > B^a$ the fraction of successes (x) over the number of trials (N), *Mean* calculated as pN where p is the probability of success (50 %), Sigma calculated as $\sqrt{Np(1-p)}$, z-stat calculated as $(x - pN)/\sqrt{Np(1-p)}$, p-value from a right tailed z-test. Results concern the whole period of examination

outperform the corresponding artificial ones in more cases than expected by true chance (Table 4.2). This is not the case for the resistance levels. A right tailed z-test is performed since we care on the cases where $B^e > B^a$ are substantial greater than 50 %. Only z-values corresponding to support levels and HSARs as a whole are positive and statistically significant at 1 % level. This suggests that price bounces more often than would be expected by true chance on HSARs, and support levels, and less often in cases of resistance levels. Furthermore, the z-values of support levels are much greater than those regarding the combined HSARs which explain the superiority of supports compared with resistances. This superiority constitutes combined HSARs as efficient trend-reversal predictors.

Osler (2000) used published support and resistance levels provided by six firms. Three of these firms provided estimated strengths at each one of their published SAR. These strengths took values from 1 to 3 with the value of 3 being the strongest. Concerning the importance of these "strengths" she states:

> In short, published estimates of the strength of the levels do not seem to be useful.

Technical analysts assert that the number of locals that generate a specific SAR is of prime importance. A relevant statement is found in the description of trendlines in Pring (2002). More concretely, he mentions:

> A trendline derives its authority from the number of times it has been touched or approached; that is, the larger the number, the greater the significance. ... a trendline represents a dynamic area of support and resistance.

Consequently, we can classify the bounce frequencies of HSARs based on the number of locals used for their identification. We could argue that we can treat them as relevant strengths of each HSAR. Results for both markets for the whole sample period are presented in Table 4.3. Taking a first look at the tabulated results, one notices that the number of locals does not seem to affect the bounce frequencies with an exception the seventh strength (8 locals) in the case of NYSE. In this case we would expect a much greater bounce frequency since 8 locals are the greatest strength of this market but actually the bounce frequency is much lower compared with the other weaker strengths. But the number of observations in this case, (and also in the case of 7 locals) is extremely low so we cannot extract any rigorous statistical conclusion from it. In other cases the number of observations is sufficiently great so we proceed with the following test.

Table 4.3 Local's effect on bounce frequencies

	ALL HSARs			Supports			Resistances		
Locals	Hits	Bounces	BF (%)	Hits	Bounces	BF (%)	Hits	Bounces	BF (%)
Panel A: NYSE									
2	58,716	35,680	60.77	29,327	18,267	62.29	29,389	17,413	59.25
3	14,954	9,006	60.22	7,714	4,801	62.24	7,240	4,205	58.08
4	3,784	2,308	60.99	1,923	1,187	61.73	1,861	1,121	60.24
5	908	531	58.48	459	276	60.13	449	255	56.79
6	146	92	63.01	70	43	61.43	76	49	64.47
7	25	17	68.00	7	4	57.14	18	13	72.22
8	10	3	30.00	6	2	33.33	4	1	25.00
Panel B: NASDAQ									
2	26,868	16,386	60.99	13,601	8,285	60.91	13,267	8,101	61.06
3	6,661	4,066	61.04	3,321	2,007	60.43	3,340	2,059	61.65
4	1,835	1,095	59.67	948	564	59.49	887	531	59.86
5	382	235	61.52	212	136	64.15	170	99	58.24
6	116	65	56.03	62	38	61.29	54	27	50.00

BF(%) bounce frequencies

Bounce frequencies generated by every possible pair of strengths are compared (Tables 4.4–4.6). Specifically, the comparison between the bounce frequencies generated by HSAR with the second and first strengths (three and two locals respectively) is symbolized as "$2 \rightarrow 1$". In case of NASDAQ, stock prices hit HSAR (irrespectively if these HSAR are support or resistance levels) 26,868 times when SARs are drawn by two locals. 16,386 of those hits are recorded as bounces, giving a bounce frequency of 60.99 %. If the strength of HSAR is meaningful, HSAR created by three or more locals should realize greater statistically significant bounce frequencies. Each hit is treated as a "Bernoulli trial" that has two possible outcomes: success (bounce) or failure. The probability of having a bounce is given by the bounce frequency of the weaker strength. In our example ($2 \rightarrow 1$), prices hit HSAR of three locals 6,661 times and bounce 4,066 times. Is this bounce frequency (61.04 %) greater and statistically significant than the one generated with "two-local" strength (60.99 %)? With 6,661 being the number of trials, N, and the bounce frequency of the weaker strength (60.99 %) being the probability of success of each trial, p, the mean, μ and the standard deviation, σ for the binomial distribution are calculated. Denote μ^s and σ^s the mean and the standard deviation calculated with the number of trials (hits) of the superior strength and a probability of success, the bounce frequency generated by the weaker strength. Subsequently we perform a right-tailed z-test of the null hypothesis that the number of bounces generated by the "strongest" strength is a random sample from a normal distribution with mean μ^s and a standard deviation of σ^s, against the alternative that the mean is greater than the μ^s. The same process is followed for all comparisons between all pairs of strengths. Furthermore, we examine the importance of strengths in isolated support and resistance levels. For brevity, only the results of NASDAQ are presented in Table 4.4. Results are aligned with the ones reported in Osler (2000). The number of locals that generate HSAR seems to be irrelevant with

Table 4.4 Testing the importance of HSARs' strengths in NASDAQ dataset

Comparison	N	p	Successes	μ^s	σ^s	z-Stat.	p-Value
$2 \rightarrow 1$	6,661	0.6099	4,066	4,062.54	39.81	0.07	0.4654
$3 \rightarrow 1$	1,835	0.6099	1,095	1,119.17	20.89	–1.16	0.8763
$4 \rightarrow 1$	382	0.6099	235	232.98	9.53	0.21	0.4162
$5 \rightarrow 1$	116	0.6099	65	70.75	5.25	–1.09	0.8631
$3 \rightarrow 2$	1,835	0.6104	1,095	1,120.08	20.89	–1.20	0.8851
$4 \rightarrow 2$	382	0.6104	235	233.17	9.53	0.19	0.4240
$5 \rightarrow 2$	116	0.6104	65	70.81	5.25	–1.11	0.8655
$4 \rightarrow 3$	382	0.5967	235	227.94	9.59	0.74	0.2307
$5 \rightarrow 3$	116	0.5967	65	69.22	5.28	–0.80	0.7876
$5 \rightarrow 4$	116	0.6152	65	71.36	5.24	–1.21	0.8877

For stocks traded at NASDAQ, it is being evaluated if the strength characterising HSAR is meaningful. Each "hit" is treated as a Bernoulli trial with two possible outcomes; success (bounce) or failure. The probability of success in each pair of comparison is the bounce frequency realized from the inferior strength. Mean and Sigma are the corresponding mean and standard deviation. z-stat and p-values result after performing a right tail z-test

Table 4.5 Testing the importance of support levels' strengths in NASDAQ dataset

Comparison	N	p	Successes	μ^s	σ^s	z-Stat.	p-Value
$2 \to 1$	3,321	0.6091	2,007	2,022.82	28.12	−0.56	0.7132
$3 \to 1$	948	0.6091	564	577.43	15.02	−0.89	0.8143
$4 \to 1$	212	0.6091	136	129.13	7.10	0.97	0.1668
$5 \to 1$	62	0.6091	38	37.76	3.84	0.06	0.4755
$3 \to 2$	948	0.6043	564	572.88	15.06	−0.59	0.7223
$4 \to 2$	212	0.6043	136	128.11	7.12	1.11	0.1339
$5 \to 2$	62	0.6043	38	37.47	3.85	0.14	0.4449
$4 \to 3$	212	0.5949	136	126.12	7.15	1.38*	0.0834
$5 \to 3$	62	0.5949	38	36.88	3.87	0.29	0.3864
$5 \to 4$	62	0.6415	38	39.77	3.78	−0.47	0.6807

As for Table 4.4

*Represents statistical significance at the 10 % level

Table 4.6 Testing the importance of resistance levels' strengths in NASDAQ dataset

Comparison	N	p	Successes	μ^s	σ^s	z-Stat.	p-Value
$2 \to 1$	3,340	0.6106	2,059	2,039.40	28.18	0.70	0.2434
$3 \to 1$	887	0.6106	531	541.60	14.52	−0.73	0.7673
$4 \to 1$	170	0.6106	99	103.80	6.36	−0.76	0.7750
$5 \to 1$	54	0.6106	27	32.97	3.58	−1.67	0.9522
$3 \to 2$	887	0.6165	531	546.84	14.48	−1.09	0.8629
$4 \to 2$	170	0.6165	99	104.81	6.34	−0.92	0.8201
$5 \to 2$	54	0.6165	27	33.29	3.57	−1.76	0.9609
$4 \to 3$	170	0.5986	99	101.76	6.39	−0.43	0.6672
$5 \to 3$	54	0.5986	27	32.32	3.60	−1.48	0.9303
$5 \to 4$	54	0.5824	27	31.45	3.62	−1.23	0.8902

As for Table 4.4

the bounce frequencies realized. This finding is aligned with the finding reported in Osler (2000) whereby published estimates of the strength of these levels did not seem to be meaningful.

4.5.2 *Profitability Analysis*

The former analysis regards the evaluation of HSARs as efficient trend reversal predictors. As presented in the Sect. 4.5.1, estimated support levels signal trend reversals more successfully than artificial ones. However, this performance is not adequate to characterize the HSARs as efficient. What is of prime importance is to answer the following question: Can this automatic identification system generate systematically, statistically significant excess returns? Even if the aforementioned system generates positive trading returns, these returns have to be compared with unconditional returns to examine the potential presence of excess returns. For this reason, in this section the corresponding profitability analysis is presented.

The returns used here are logarithmic.[7] After calculating the logarithmic returns produced by the trading rules, a comparison is implemented with the market average returns in a manner similar with that used in Leigh et al. (2004) and Wang and Chan (2007, 2009).

$$\overline{R}_{M,q} = \frac{\sum_{t=m}^{n} \ln\left(p_{t+q}/p_t\right)}{n-m+1} \tag{4.18}$$

Here, $\overline{R}_{M,q}$ is the mean market return when using a constant holding period of q, p_t is the adjusted closing price on trading day t and finally m (n) is the first (last) trading day in the subperiod of comparison. Actually, the terms in the summation of numerator of (4.18) form different distributions of market returns for every single holding period (q), in a sense that long positions are taken every day and are held for q days. The excess mean return is calculated by subtracting the mean market return from the mean trading rule return. Since we care for the statistical significance of positive excess returns, the comparison between the mean market return and the mean trading rule return is managed by a two-sample, one-tailed, unequal variance (heteroskedastic) Student's t-test. Trade returns generated by initial short positions are analogously compared with the market returns generated by short selling and holding for the aforementioned sets of holding periods.

Figures 4.9 and 4.10 consists of four binary images which illustrate the cases where a given stock for a given holding period realizes statistically significant excess returns, for a significance level of 5 %. The vertical axis of each sub-image corresponds to the stocks and the horizontal axis to the 30 different holding periods considered.

Black dots consist the minority so the first indication we take is that at most stocks, the trades signaled by HSARs cannot outperform a simple buy and hold strategy. In case of NASDAQ resistance levels, compared with support levels, seem to generate statistically significant excess returns in more stocks. The opposite holds for NYSE stocks, but generally the differences between supports and resistance levels' profitability are commensurate nuance. In cases where the black dots are characterized by continuity, horizontal black lines appear. These rare cases correspond to the stocks where HSAR predict major trend reversals which last for relative long period of time. The picture for TR_2 is exacerbated (Fig. 4.10) as expected. The reason is that, since the bounce analysis showed efficiency in the predictability of trend reversals we could expect *a priori* the first trading rule to have more chances in producing systematically excess returns. Results presented in this section, confirm this, but even the superiority of the first trading rule is not sufficient enough to characterize the examined markets as weak form inefficient.

---

[7] The results reported in this part are robust, and not affected with a use of arithmetic returns.

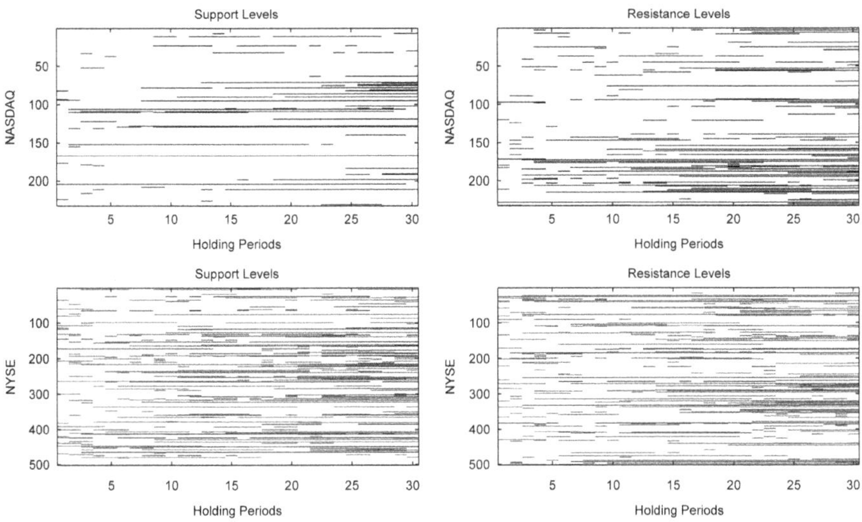

Fig. 4.9 Spotting abnormal returns for TR_1.The *black points* in the above binary images indicate the cases where a given stock for a given holding period generates statistically significant excess returns for a significance level of 5 %. The test implemented is a two-sample, one-tailed, unequal variance (heteroscedastic). Student's *t*-test. *Left* (*right*) sub-images correspond to support (resistance) levels, and *upper* (*lower*) sub-images regard NASDAQ (NYSE) stock market. The results are for the first trading rule (TR_1) and for the whole sample period

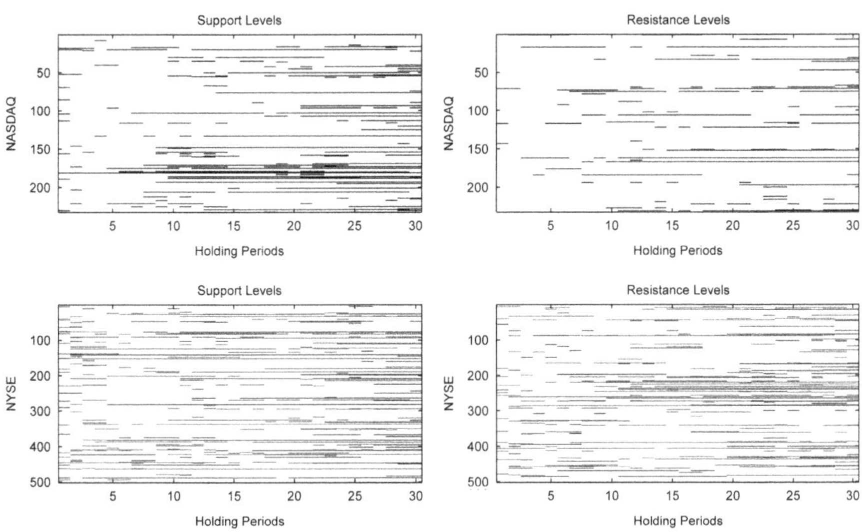

Fig. 4.10 Spotting abnormal returns for TR_2

The same analysis is implemented for all sub-periods of comparison, and the results are summarized in Figs. 4.11 and 4.12. Concerning support levels a mild improved performance can be seen in the second subperiod of comparison in longer term holding periods. Generally the number of stocks that generate excess returns is dismally low, in both markets and for all sub-periods of comparison, proportionally to the initial size of our sample.

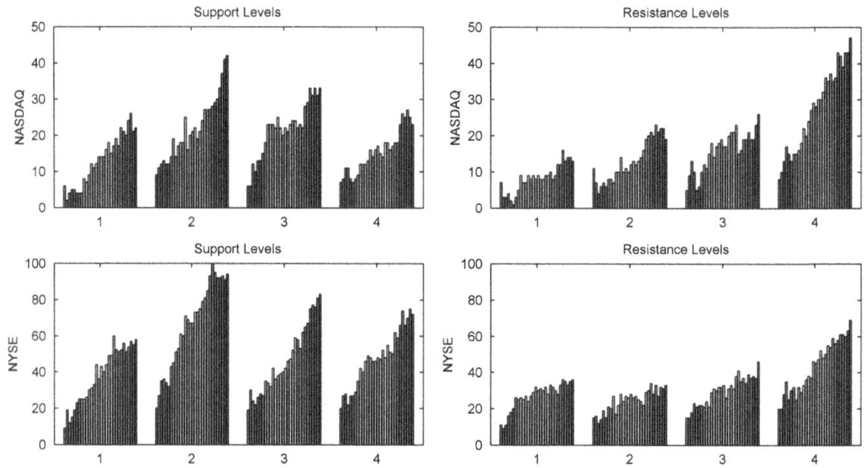

Fig. 4.11 The breakdown of the profitability to sub-periods of comparison (TR_1). These plots illustrate the number of stocks that present statistically significant excess returns (at a significance level of 5 %), when the first trading rule (TR_1) is used, for the various sub-periods. 1, 2 and 3 sets of bars correspond to the three sub-periods of comparison and the fourth to the whole sample period. Each set of bars includes 30 bars for every different holding period

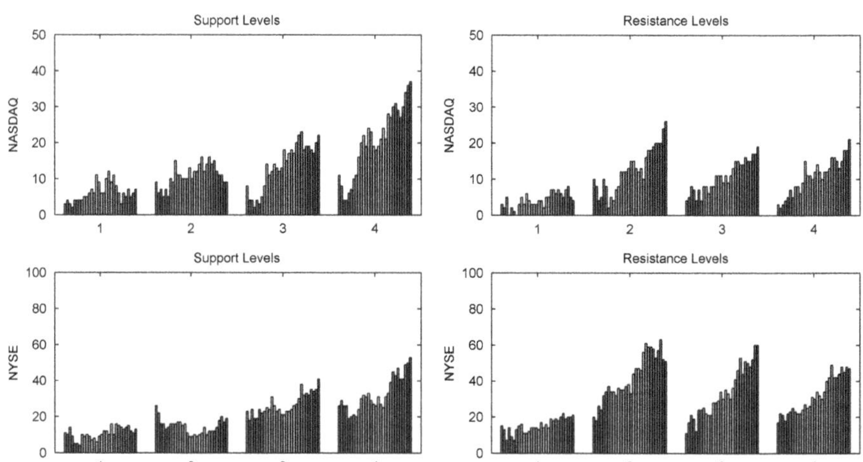

Fig. 4.12 The breakdown of the profitability to sub-periods of comparison for the second trading rule (TR_2)

4.5.3 *Consistency*

Results reported in Sect. 4.5.2 are not enough to characterize HSAR as inefficient. Is it possible to "beat" the market if we isolate the stocks that realize statistically significant excess returns in preceding subperiods? In other words, after running the simulation for the first subperiod, would the "profitable" stocks be also "profitable" in the succeeding subperiod? In Zapranis and Tsinaslanidis (2012a) we have found that there was no consistency in the profitability of HSARs on subsequent subperiods. More precisely, there were up to two stocks in each dataset where HSARs exhibited consistent profitability for two subsequent subperiods, whilst HSARs did not present significant excess returns in all three subperiods in any of the stocks considered.

4.6 Conclusions

The proposed HSAR algorithmic method encloses principles found in the literature for the visual identification of SARs. Since most other articles identify support (resistance) levels as the previous lowest (greatest) price, the comparison of our findings was made with those presented in Curcio et al. (1997) and Osler (2000). The reason that these two studies are used as benchmark, is that the evaluated support and resistance levels had been published by proficient firms rather than calculated by simple arithmetic rules.

After identifying successfully HSARs, it was assessed whether these levels were efficient trend-reversal predictors (Sect. 4.5.1), and if they could generate systematically abnormal returns (Sect. 4.5.2). Results presented corroborate the most basic conclusions reported in the aforementioned studies of comparison. Regarding the "bounce analysis", stock prices of the examined dataset bounced on estimated support levels more often than would have occurred had the HSAR been randomly chosen. This constitutes support levels as efficient trend reversal predictors. On the other hand, estimated resistance levels failed to outperform the artificial ones. The superiority of support levels compensates for the poorer performance of resistance levels, and this affects the overall picture of HSAR which was found to be generally efficient.

Furthermore, it has been examined if the number of locals that signal a HSAR has any importance. The number of these locals is treated as a strength value for each HSAR and results indicate that these "strengths" play no major role in predicting trend interruptions. "Bounce analysis" confirms some of the main findings reported in Osler (2000) for currency markets, with the difference that this study scrutinizes further the performance of the support and resistance levels separately.

In the part of "profitability analysis", two different trading rules are used: trading after bounces (TR_1) and trading after penetrations (TR_2). The first trading rule, outperforms the second one, as expected, but still, without even considering transactions costs, fails to generate systematically statistically significant excess returns. This finding confirms the results presented in Curcio et al. (1997). Furthermore no consistency in the profitable performance is found in the examined stock price series, which means that stocks that generate abnormal returns fail to "beat the market" across subsequent subperiods. In an effort to explain the aforementioned results, the better performance of support levels can be attributed to the generally long term "bullish" trend that characterizes the stocks' price series considered. More precisely, a positive trend has two main impacts: First, price has more (less) chances to bounce on support (resistance) levels rather than breach them. Second, it generates substantially high mean unconditional returns which are greater than the positive trading returns resulted by HSARs. Thus, the performance concerning the trend predictability is somehow elusive since the profitability tests implemented fail to provide evidence of abnormal returns compared with a simple buy-and-hold strategy.

Appendix 1: HSAR Function

1.1. The Function

$$[SAR, Bounds, Freq, x\_act] = HSAR(ys, w, x, pflag)$$

1.2. Description

The $HSAR(\cdot)$ identifies Horizontal Support and Resistance Levels (HSARs) at time t conditioning information up to time $t-1$. $RW(\cdot)$ is embedded in this function.

Inputs:

1. *ys*: Price series (column vector).
2. *w*: Is used to define the size of the rolling window which is 2w + 1 (see $RW(\cdot)$ in Appendix 1 of Chap. 2)
3. *x*: The desired percentage that will give the bounds for the HSARs (e.g. 5 %).
4. *pflag*: If 1 the function returns two plots.

Outputs

1. *SAR*: Horizontal Support and resistance levels
2. *Bounds*: Bounds of bins used to classify the peaks and bottoms.
3. *Freq*: Frequencies for each bin
4. *x_act*: Actual percentage of the bins' distance

1.3. Code

```
function [SAR,Bounds,Freq,x_act]=HSAR(ys,w,x,pflag)
%% HSARs Identification
[Peaks,Bottoms]=RW(ys,w,0);    % Identify regional locals
L=[Peaks(:,1);Bottoms(:,1)];    % All locals
L1=min(L)/(1+x/2);          % Lower bound
Ln=max(L)*(1+x/2);          % Upper bound
n=(log(Ln/L1))/(log(1+x));     % Number of bins (aproximation)
N_act=round(n);            % Actual number of bins
x_act=nthroot(Ln/L1,N_act)-1;   % Actual percentage for each bin
Bounds=L1*(1+x_act).^(0:N_act)'; % Bounds for bins
Freq=zeros(N_act,1);
for i=1:N_act
    Freq(i,1)=sum( L(:,1)>=Bounds(i,1) & L(:,1)<Bounds(i+1,1));
end % Frequencies per bin
idx=find(Freq>=2);
if isempty(idx)==1
    disp('No HSARs identified')
    SAR=[];
else
    m=length(idx);                  % Number of HSARs
    SAR=(Bounds(idx,:)+Bounds(idx+1,:))/2;
    %% Plotting
    if pflag==1
        Pc=cell(m,1);Bc=Pc;
    for i=1:m
        Pc{i,1}=find(Peaks(:,1)>=Bounds(idx(i),1) &...
                Peaks(:,1)<=Bounds(idx(i)+1,1));
            Pyx=Peaks(cell2mat(Pc),:);
            Bc{i,1}=find(Bottoms(:,1)>=Bounds(idx(i),1) &...
                Bottoms(:,1)<=Bounds(idx(i)+1,1));
            Byx=Bottoms(cell2mat(Bc),:);
        end
        figure(1) % Price series with HSARs and Locals
        plot(ys),hold on
        t=[1,length(ys)];
        for i=1:m
            plot(t,[SAR(i),SAR(i)],'r')
        end
        plot(Pyx(:,2),Pyx(:,1),'r*')
        plot(Byx(:,2),Byx(:,1),'ro')
        hold off
```

```
        figure(2)% Histogram
        hold on
        for i=1:N_act
            bar((Bounds(i,1)+Bounds(i+1,1))/2,...
                Freq(i,1),Bounds(i+1,1)-(Bounds(i,1)))
        end
        xlabel('Price'),ylabel('Frequency'),hold off
    end
end
```

Appendix 2: HSARsim Function

2.1. The Function

$$[Results] = HSARsim(ys, w, x, indata, pflag)$$

2.2. Description

The $HSARsim(\cdot)$ identifies Horizontal Support and Resistance Levels (HSARs) on a daily basis. $RW(\cdot)$ and $HSAR(\cdot)$ are embedded in this function.

Inputs:

1. *ys*, *w*, and *x*: Same description with $HSAR(\cdot)$.
2. *indata*: Is an integer number defining the size of the initial subsample used for the identification of the first HSARs.
3. *pflag*: If 1 the function returns the corresponding plot.

Outputs

1. *Results*: Structure array containing the results of the function. Particularly, if l is the number of observations of the examined price series and n is the maximum number of HSARs identified on a specific day then,
   - *Results.NofSARs*: Is a $1 \times l$ vector containing the number of HSARs identified on a daily basis
   - *Results.x_act*: Is a $1 \times l$ vector containing the corresponding *x_act* values returned by $HSAR(\cdot)$ on a daily basis.
   - *Results.ActSARs*: Is a $n \times l$ matrix containing the HSARs identified for every day.
   - *Results.PofSARs*: Is a $n \times l$ matrix containing the number of regional locals used for the identification of HSARs presented in *Results.ActSARs*. Technical

analysts assert that these figures are of prime importance and thus we can treat them as a relative strength of the estimated HSARs.

When for a particular day no HSAR is identified corresponding cells of the above outputs take the value of nan.

2.3. Code

```
function [Results]=HSARsim(ys,w,x,indata,pflag)
l=length(ys);
Results.NofSARs=nan(1,l);Results.x_act=nan(1,l);
Results.ActSARs=nan(100,l);Results.PofSARs=nan(100,l);
for i=indata:l-1
    [SAR,~,Freq,x_act]=HSAR(ys(1:i),w,x,0);
    NofSARs=size(SAR,1);     % Number of HSARs
    if NofSARs>0
        PofSARs=Freq((Freq>=2)); % Power of HSARs
  else SAR=nan;
        PofSARs=nan;
  end
  Results.NofSARs(1,i+1)=NofSARs;
  Results.x_act(1,i+1)=x_act;
  a=100-NofSARs;
  if NofSARs>0
        a=nan(a,1);
  else a=nan(100-1,1);
  end
  Results.PofSARs(:,i+1)=[PofSARs;a];
  Results.ActSARs(:,i+1)=[SAR;a];
end
Results.ActSARs(max(Results.NofSARs)+1:100,:)=[];
Results.PofSARs(max(Results.NofSARs)+1:100,:)=[];
%% Plotting
if pflag==1
    bu=ys'*1.2;bu=[nan bu(2:end)];
    bu=repmat(bu,size(Results.ActSARs,1),1);
    bd=ys'*0.8;bd=[nan bd(2:end)];
    bd=repmat(bd,size(Results.ActSARs,1),1);
    idx1=Results.ActSARs>bu;
    idx2=Results.ActSARs<bd;
    PosSARs=Results.ActSARs;
    PosSARs(idx1)=nan;
    PosSARs(idx2)=nan;
```

```
    figure(1)
    plot(ys);hold on
    plot(PosSARs','k.');
    xlabel('Date'),ylabel('Price')
    xlim([1,length(ys)])
    hold off
    set(gca,'LooseInset',get(gca,'TightInset'))
end
end
```

References

Achelis SB (1995) Technical analysis from A to Z. Probus Publishing, Chicago

Bessembinder H, Chan K (1995) The profitability of technical trading rules in the Asian stock markets. Pac Basin Financ J 3:257–284

Brock W, Lakonishok J, Lebaron B (1992) Simple technical trading rules and the stochastic properties of stock returns. J Financ 48(5):1731–1764

Bulkowski TN (2002) Trading classic chart patterns. Wiley, New York

Cheung W, Lam KSK, Yeung H (2011) Intertemporal profitability and the stability of technical analysis: evidences from the Hong Kong stock exchange. Appl Econ 43:1945–1963

Chong TT-L, Ng W-K (2008) Technical analysis and the London stock exchange: testing the MACD and RSI rules using the FT30. Appl Econ Lett 15:1111–1114

Coutts JA, Cheung K-C (2000) Trading rules and stock returns: some preliminary short run evidence from the Hang Seng 1985-1997. Appl Financ Econ 10:579–586

Curcio R, Goodhart C, Guillaume D, Payne R (1997) Do technical trading rules generate profits? Conclusions from the intra-day foreign exchange market. Int J Financ Econ 2:267–280

De Grauwe P, Decupere D (1992) Psychological barriers in the foreign exchange market. J Int Comp Econ 1:87–101

De Zwart G, Markwat T, Swinkels L, van Dijk D (2009) The economic value of fundamental and technical information in emerging currency markets. J Int Money Financ 28:581–604

Dorfleitner G, Klein C (2009) Psychological barriers in European stock markets: where are they? Global Financ J 19(3):268–285

Edwards RD, Magee J (1997) Technical analysis of stock trends, 7th edn. John Magee, Inc., Boston

Falbo P, Cristian P (2011) Stable classes of technical trading rules. Appl Econ 43:1769–1785

Goodhart C, Figliuoli L (1991) Every minute counts in financial markets. J Int Money Financ 10 (1):23–52

Hudson R, Dempsey M, Keasey K (1996) A note on weak form efficiency of capital markets: the application of simple technical trading rules to UK prices 1935–1994. J Bank Financ 20:1121–1132

Kavajecz KA, Odders-White ER (2004) Technical analysis and liquidity provision. Rev Financ Stud 17(4):1043–1071

Leigh W, Modani N, Hightower R (2004) A computational implementation of stock charting: abrupt volume increase as signal for movement in New York Stock Exchange Composite Index. Decis Support Syst 37:515–530

Lo AW, Mamaysky H, Wang J (2000) Foundations of technical analysis: computational algorithms, statistical inference, and empirical implementation. J Financ 55(4):1705–1765

Marshall BR, Cahan RH, Cahan JM (2008a) Can commodity futures be profitably traded with quantitative market timing strategies? J Bank Financ 32:1810–1819

Marshall BR, Cahan RH, Cahan JM (2008b) Does intraday technical analysis in the U.S. equity market have value? J Empir Financ 15:199–210

Marshall BR, Qian S, Young M (2009) Is technical analysis profitable on US stocks with certain size, liquidity or industry characteristics? Appl Financ Econ 19:1213–1221

Mitchell J (2001) Clustering and psychological barriers: the importance of numbers. J Futur Mark 21(5):395–428

Mitchell J (2006) Clustering and psychological barriers in exchange rates. J Int Financ Mark Inst Money 16(4):318–344

Mizrach B, Weerts S (2009) Highs and lows: a behavioural and technical analysis. Appl Financ Econ 19:767–777

Murphy J (1986) Technical analysis of the future markets: a comprehensive guide to trading methods and applications. Prentice Hall, New York

Osler CL (2000) Support for resistance: technical analysis and intraday exchange rates. FRBNY Econ Policy Rev 6(2):53–68

Pring MJ (2002) Technical analysis explained: the successful investor's guide to spotting investment trends and turning points, 4th edn. McGraw-Hill Book Company, New York

Qi M, Wu Y (2006) Technical trading-rule profitability, data snooping, and reality check: evidence from the foreign exchange market. J Money Credit Bank 38(8):2135–2158

Raj M, Thurston D (1996) Effectiveness of simple technical trading rules in the Hong Kong futures markets. Appl Econ Lett 3:33–36

Wang J, Chan S (2007) Stock market trading rule discovery using pattern recognition and technical analysis. Expert Syst Appl 33:304–315

Wang J, Chan S (2009) Trading rule discovery in the US stock market: an empirical study. Expert Syst Appl 36:5450–5455

Zapranis A, Tsinaslanidis PE (2012a) Identifying and evaluating horizontal support and resistance levels: an empirical study on US stock markets. Appl Financ Econ 22:1571–1585. doi:10.1080/09603107.2012.663469

Zapranis A, Tsinaslanidis PE (2012b) A novel, rule-based technical pattern identification mechanism: Identifying and evaluating saucers and resistant levels in the US stock market. Expert Syst Appl 39(7):6301–6308. doi:10.1016/j.eswa.2011.11.079

Chapter 5
Zigzag Patterns

5.1 Introduction

As implied by their name, zigzag patterns generally appear when price evolves sharply in alternating directions. This price behaviour generates sequences of local extrema (regional peaks and troughs). The majority of technical price patterns lie in this category. Therefore, in terms of the number of patterns examined, this is the more extensive chapter of the book. The importance of zigzag patterns is highlighted in Sklarew (1980):

> The zigzag pattern is the foundation of all chart formations, and is the key to their forecasting value. ... All chart formations are some variation of either the minor or major zigzag trend pattern.

Aligned is the following assertion found in Neftci (1991):

> ...most patterns used by technical analysts need to be characterized by appropriate sequences of local minima and/or maxima....

From the bibliography dealing with technical patterns, the majority focuses on zigzag patterns. Although not exhaustive, remarkable are the following studies which we recommend the reader to consider: (Levy 1971; Osler and Chang 1995; Clyde and Osler 1997; Chang and Osler 1999; Lo et al. 2000; Dawson and Steeley 2003; Lucke 2003). Important empirical findings and conclusions of these studies have already been presented in Chap. 1.

In this chapter, six well-known zigzag patterns will be examined. These patterns are the *head and shoulders* (HS), *triple tops* (TT) and *triple bottoms* (TB), *double tops* (DT) and *double bottoms* (DB), flags, pennants and wedges. Algorithmic recognizers for the identification of the aforementioned patterns will be developed and presented. The recognition process requires the prior identification of regional peaks and troughs. Subsequently it will be assessed whether subsequences of these local extrema fulfill particular condition sets in order to confirm a pattern's fully formation. These quantitative condition sets result from theoretical descriptions of

P.E. Tsinaslanidis, A.D. Zapranis, *Technical Analysis for Algorithmic Pattern Recognition*, DOI 10.1007/978-3-319-23636-0_5

the examined patterns found in various technical manuals. The proposed algorithms will be adopted on real and simulated datasets. The first dataset consists of NASDAQ and NYSE components whereas the second one consists of simulated series generated by the discrete-time version of *geometric Brownian motion* (GBM). The price behaviour after each pattern's confirmation will be assessed and a comparison between results obtained from actual and simulated series will be carried. The spirit of the proposed identification mechanisms is similar across the patterns we examine and the reader will be able to develop his own recognizers for other zigzag patterns not considered in this chapter.

The rest of the chapter is organized as follows. In Sect. 5.2, the HS pattern identification mechanism is presented. Section 5.2.1, examines the possibility of identifying the HS pattern in simulated series generated by the discrete-time version of GBM. After ensuring that this technical pattern can be successfully created in a randomly generated series, it is examined how different widths of rolling windows (w) and different values of drift (μ) and volatility (σ) rates affect the number of identified patterns. In Sect. 5.3, the algorithmic methodology for the identifications of TTTB and DTDB patterns is developed. In Sect. 5.4, the proposed recognizers for flags, pennants and wedges are presented. In Sect. 5.5, the choice of w is discussed. In Sect. 5.6, trading rules are designed and three mutually exclusive price evolutions following a pattern's fully formation are identified. These distinct price behaviors are statistically assessed in Sect. 5.7, where a comparison between results obtained from real and artificial created series is made. Finally, in Sect. 5.8 we conclude. In Appendices 1 and 2 the developed Matlab codes for a selection of the examined zigzag patterns is presented.[1]

5.2 Identifying the Head and Shoulders Pattern

In technical analyst's community, the HS pattern is perhaps the most celebrated technical pattern. This is mainly attributed to the fact that technicians view it as a pattern of high credibility (Bulkowski 2000). Like most technical patterns, its name comes from the similarity of its structure with ordinary objects, to which the human mind is familiar with. More precisely, its name stems from its resemblance to the upper part of human body. This pattern has been described extensively in many books (Achelis 1995; Edwards and Magee 1997; Murphy 1999; Bulkowski 2000, 2002; Pring 2002), scientific articles (Osler and Chang 1995; Clyde and Osler 1997;

[1] For brevity reasons, we provide the proposed matlab code for the identification of the HS and flags patterns. The identification of TTTB and DTDB is very similar with that of the HS whilst the identification of pennants and wedges is similar to that of flags. The reader should use these codes as examples of how to implement the identification of various zigzag patterns. We do not advise to use them for any trading purposes. The user should carefully assess their predictive performance and make any necessary amendments for further improvements.

Chang and Osler 1999; Lucke 2003; Zapranis and Tsinaslanidis 2010a, b) and online resources.

It can be observed in its normal (HS tops) or inverse (HS bottoms) form. Regarding the normal form, the pattern consists of three peaks with two intervening troughs between the head and each shoulder. The second peak observed (head), is higher than the preceding and forthcoming peaks (shoulders). The neckline is drawn through the lowest points of these two intervening troughs and may slope upwards or downwards. According to TA when the price breaks downwards the neckline, the pattern is confirmed and investors can expect a fall of the price equal to the head's height, which is the vertical distance between the head and the neckline. Chartists also assert that after the pattern's confirmation, the neckline behaves as a resistance level. The inverse form is actually a mirror image of the normal form. In terms of trading recommendation, HS tops and HS bottoms provide short and long trading signals respectively. Since they are considered generally as trend reversal patterns, we expect a realization of an uptrend (a downtrend) before the HS tops (bottom) formation. However, in case the reverse stands, these patterns return the same trading signals but they are characterized as trend continuation patterns (Murphy 1999). A simplified illustration of the HS tops pattern with its characteristic components is presented in Fig. 5.1. We use the same notation presented in Lucke (2003) to symbolize regional peaks and troughs as P_i and T_i respectively, with i being an integer used to number the regional locals. The point where prices breach the neckline, and confirm the pattern's fully formation is symbolized with B.

Lucke (2003) adopts the following five conditions to identify the HS tops pattern. These conditions were initially defined in Osler and Chang (1995):

Condition 1: The head must be higher than the shoulders:

$$P_2 > \max(P_1, P_3) \tag{5.1}$$

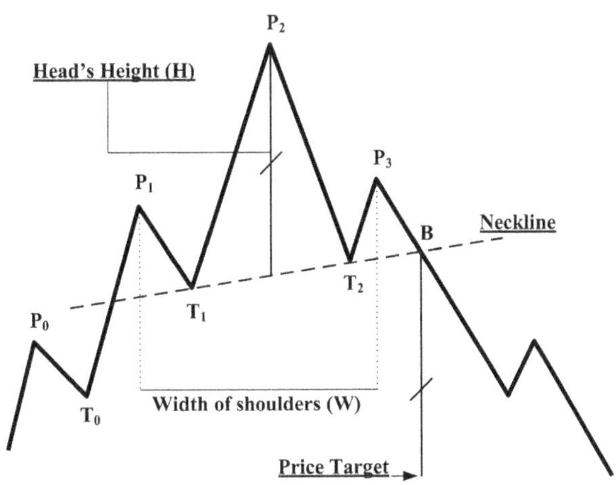

Fig. 5.1 Simplified illustration of the HS tops pattern. Pattern's characteristic components: Local peaks (points P_0, P_1, P_2 and P_3), local troughs (points T_0, T_1 and T_2), the neckline drawn by points T_1 and T_2, the head's height (distance H expressed in price), the shoulders' width (distance W expressed in days), the point of penetration (point B) and the price target set by the pattern

Condition 2. Trend preexistence: If the first set of inequalities (5.2) holds the HS tops pattern is characterized as a trend reversal pattern. Reversely, fulfillment of (5.3) implies a continuation pattern.[2]

$$P_1 > P_0 \,\&\, T_1 > T_0 \tag{5.2}$$

$$P_1 < P_0 \,\&\, T_1 < T_0 \tag{5.3}$$

Condition 3. Balance: The left (right) shoulder must be at least as high as the midpoint between right (left) shoulder and its preceding (following) trough.

$$P_1 \geq 0.5(P_3 + T_2) \,\&\, P_3 \geq 0.5(P_1 + T_1) \tag{5.4}$$

Condition 4. Symmetry: The time between left shoulder and head must not be more than 2.5 times the time between the head and right shoulder and vice versa.

$$t_{P_2} - t_{P_1} < 2.5(t_{P_3} - t_{P_2}) \,\&\, t_{P_3} - t_{P_2} < 2.5(t_{P_2} - t_{P_1}) \tag{5.5}$$

Condition 5. Penetration: Recall that the pattern is confirmed when price breaches downwards the neckline (5.6) for the first time. The time required for this confirmation, t_B, must be no longer than the time interval between the two shoulders (5.7).

$$B < \frac{T_2 - T_1}{t_{T_2} - t_{T_1}}(t_B - t_{T_1}) + T_1 \tag{5.6}$$

$$t_B < t_{P_3} + (t_{P_3} - t_{P_1}) \tag{5.7}$$

The former criteria are crucial since they combine at least four technical concepts; smoothed trends, trend reversal, resistance levels, and volatility clustering (Lucke 2003). All the aforesaid criteria are inequalities that include local peaks and troughs identified by $RW(\cdot)$.[3] One way to identify the HS, is to construct a 7 × 7 binary matrix where rows refer to the locals sorted in ascending order by the time of observation and columns correspond to the 5 criteria. When a regional local satisfies the corresponding criteria it takes the value of one. Otherwise it takes the value of zero. For a confirmed pattern a diagonal of one's is needed like the one presented in Table **5.1**. This process is used in Zapranis and Tsinaslanidis (2010a, b).

The HS bottoms price pattern (Fig. **5.2**) is a mirror image of the HS tops. Reversed is also its trading recommendation since its confirmation signals a long

[2] Osler and Chang (1995) and Lucke (2003) focus on the identification of the HS tops reversal pattern and thus, they use only the first set of inequalities.

[3] For the identification of local extrema, Lucke (2003) used a computer program (Bry and Boschan 1971) originally designed to identify business cycle turning points. PIPs presented in Chap. 2 can be used alternatively.

Table 5.1 HS tops identification on Archer Daniels Midland co. (ADM) stock

Day	P_0	T_0	P_1	T_1	P_2	T_2	P_3	Price
4	1	0	0	0	0	0	0	21.69
60	0	1	0	0	0	0	0	16.31
123	0	0	1	0	0	0	0	22.57
154	0	0	0	1	0	0	0	19.59
189	0	0	0	0	1	0	0	22.98
234	0	0	0	0	0	1	0	19.98
296	0	0	0	0	0	0	1	22.94

Compare with Figs. 5.1 and **5.3**

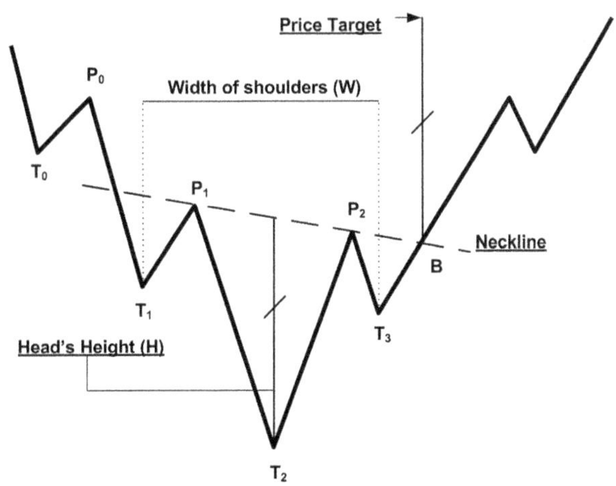

Fig. 5.2 Simplified illustration of the HS bottoms pattern

position. Similarly to HS tops, the following five conditions must be fulfilled for the successful recognition of HS bottoms:

Condition 1: The head must be lower than the shoulders:

$$T_2 > \min(T_1, T_3) \tag{5.8}$$

Condition 2. Trend preexistence: (5.2) and (5.3) are also used in this case. The difference here is that when (5.2) holds the HS bottoms is characterized as a trend continuation pattern, whereas (5.3) classifies the pattern as a trend reversal.

Condition 3. Balance:

$$T_1 \leq 0.5(T_3 + P_2) \;\&\; T_3 \leq 0.5(T_1 + P_1) \tag{5.9}$$

Condition 4. Symmetry:

$$t_{T_2} - t_{T_1} < 2.5(t_{T_3} - t_{T_2}) \;\&\; t_{T_3} - t_{T_2} < 2.5(t_{T_2} - t_{T_1}) \tag{5.10}$$

Condition 5. Penetration:

$$B > \frac{P_2 - P_1}{t_{P_2} - t_{P_1}}(t_B - t_{P_1}) + P_1 \tag{5.11}$$

$$t_B < t_{T_3} + (t_{T_3} - t_{T_1}) \tag{5.12}$$

The aforementioned conditions are embedded in the developed $HS(\cdot)$, which identifies the HS tops and bottoms simultaneously. This function takes three input variables: *ys* is the price series under consideration, *w* defines the length of the RW and *pflag* is a binary value whereby the script returns a graph if it takes the value of one. The output variable contains all necessary information for the identified patterns (for a more detailed description of this function see Appendix 1).

$$[PATTERNS] = HS(ys, w, pflag)$$

The characterization of the pattern as a trend reversal or trend continuation is rather academic, since HS tops signal short positions and HS bottoms signal long positions regardless the trend preexistence. For this reason, the developed script focuses solely on the pattern's form (normal or inverse), ignoring the second identification criteria, (5.2) and (5.3), and implementing only the four remaining ones. However it is plausible, for the user to examine further the performance of the HS pattern by breaking down the successfully identified cases into trend reversals and trend continuations. For this reason the trend preexistence should be considered, and this can be accomplished by various ways. First, the trend preexistence condition presented in (5.2) and (5.3) can be used. Second, we can characterize the preexistence by implementing a moving average[4] of d days on a specific number of days d^* preceding the pattern's initial local (P_1 for HS tops and T_1 for HS bottoms). Subsequently by dividing the number of days which are above the moving average over d^* results in a ratio that takes values in the closed interval $[0, 1]$. We can arbitrarily consider the trend as bullish (bearish) when this ratio takes values above (below) 0.5.[5]

Finally a linear polynomial fit applied on the d^* days that precede the pattern can be used to characterize the preceding trend. A polynomial model with a degree of one is actually a linear function of type

$$f(x) = bx + a, \tag{5.13}$$

[4] For this purpose the reader may use the $tsmovavg(\cdot)$ function of the financial toolbox of Matlab 2014a.

[5] Actually under the concept of TA there are three different types of trends: Uptrend (or bullish trend), downtrend (or bearish trend) and a sideways price movement (or flat trend). On this concept the aforesaid ratio can categorize the trend preexistence in one out of those three classes. In other words, the trend can be characterized as bullish, bearish or flat when the ratio takes values above 0.7, below 0.3 and between 0.3 and 0.7 respectively.

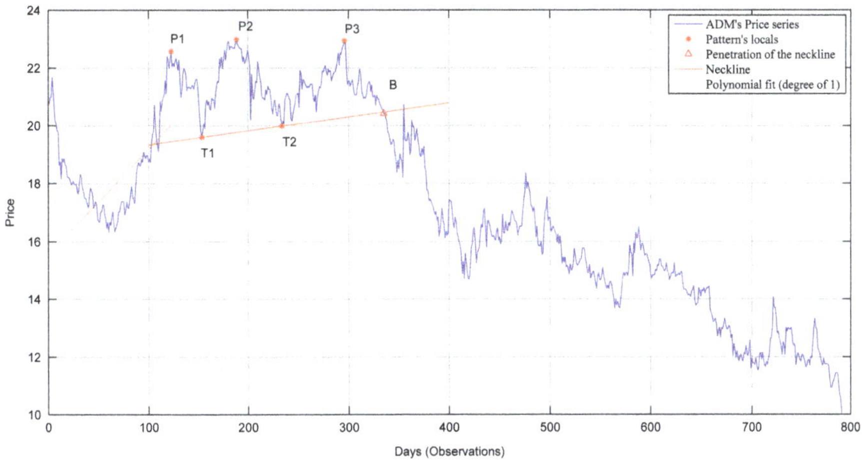

Fig. 5.3 HS tops identification on ADM stock

where b is the slope of the line and a is the intersection of the line with the vertical axis. Actually, this line can be treated as a regression trend line. Positive (negative) values of the slope indicate a bullish (bearish) trend.

In Fig. **5.3**, an example of a HS tops identification on ADM stock is presented. Variables a and b of the polynomial model where estimated on the preceding 100 days of the HS pattern $\left(d^{*}=100\right)$. Particularly, $b = 0.03641$ and $a = 15.54$. implying that the precedent trend is bullish. Table **5.1** presented earlier corresponds to the same pattern as well. The example of ADM was not randomly selected. The visual identification of this pattern is presented in http://stockcharts.com.[6] The successful algorithmic identification of this particular pattern is a first indication that the developed methodology can successfully identify cases of HS, previously identified visually.[7]

5.2.1 A Simulation Experiment

Before proceeding with the assessment of the HS pattern's predictive performance on real financial data, it is essential to examine first whether the pattern can be formed on stochastically generated price series and, second, if so, assess its performance. Should we take for granted that the pattern can be identified in a random series? The answer should be negative. TA asserts that patterns occur due to

[6] http://stockcharts.com/school/doku.php?id=chart_school:chart_analysis:chart_patterns:head_and_shoulders_top_reversal (Accessed 14 July 2014)

[7] The proposed algorithm was also tested, successfully, on several other cases found in online sources, where the HS was identified visually on stock price series.

a recursive investors' trading behavior which is not random. In other words, market participants react in a consistent manner under similar market conditions, shifting demand and supply lines accordingly and thus prices evolve in particular patterns.

The simulated dataset is constructed by using the discrete-time version of geometric Brownian motion (GBM)

$$\frac{\Delta P}{P} = \mu \Delta t + \sigma \epsilon \sqrt{\Delta t},\ \epsilon \sim N(0,1). \tag{5.14}$$

In (5.14), ΔP is the change in the stock price, P, in a small time interval Δt, μ is the expected rate of return per unit of time, σ is the volatility of the stock price and ϵ is randomly drawn from a standard normal distribution (Hull 2006). Equation (5.14) implies that $\Delta P/P$ follows a normal distribution with mean $\mu \Delta t$ and standard deviation $\sigma\sqrt{\Delta t}$. Figures **5.4** and **5.5** illustrate four simulated price paths of 500 observations using (5.14) with four different values for μ and σ respectively, *ceteris paribus*. Apparently, increasing the value of μ shifts the simulated price path upwards (Fig. 5.4), whilst prices volatile more when increasing the value of σ (Fig. 5.5).

Using (5.14) and setting an initial price of ten, 100,000 simulated price series of 1000 observations each were generated. Parameters μ and σ took ten equally distant values within the closed price intervals $[-0.5, 0.5]$ and $[0.2, 0.6]$ respectively. Parameter w which defines the size of the rolling window in $RW(\cdot)$ took five equally distant values within the closed price interval [20, 60]. For every combination of μ and σ, 1000 simulated price series were generated. A seed had been used in order to ensure that the Matlab pseudorandom generator produced the same sequences of random variables ϵ (5.14). Thus data-subsets of 1000 series differ only in respect of drift and volatility rate. Subsequently, $HS(\cdot)$ was implemented five times on each

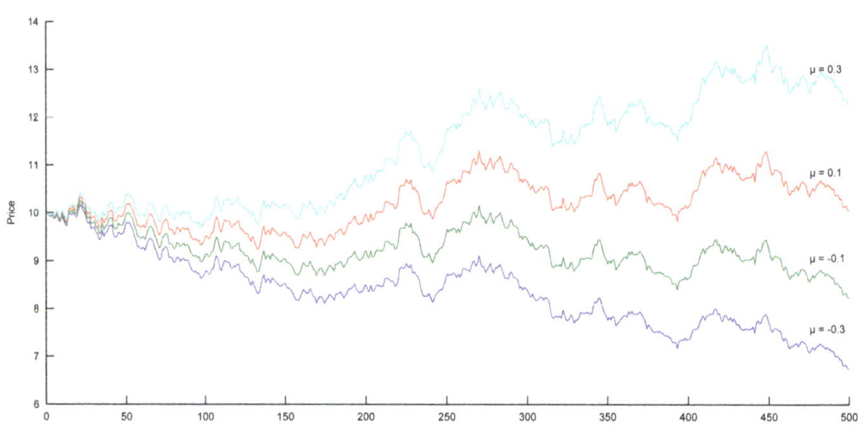

Fig. 5.4 Four simulated price paths of 500 observations using the discrete version of GBM with an initial price of 10, $\sigma = 0.2$ and $\mu = -0.3,\ -0.1$, 0.1 and 0.3. A seed was used in order the pseudorandom generator to produce the same random variables, ϵ, for all four series

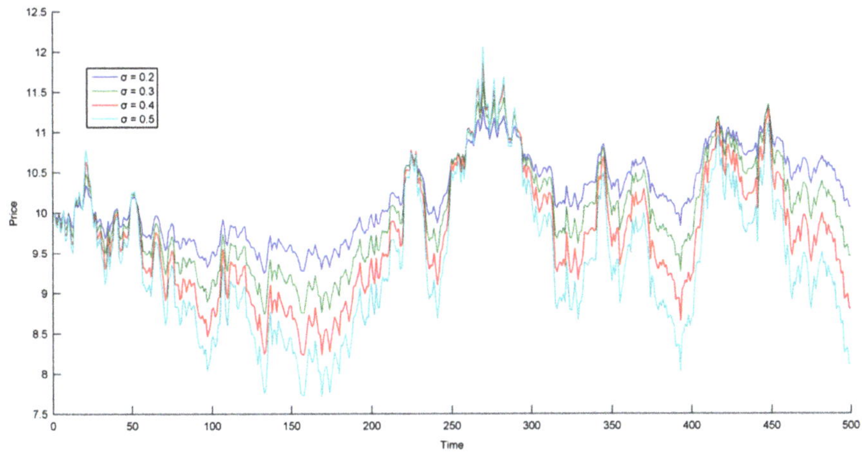

Fig. 5.5 Four simulated price paths of 500 observations using the discrete version of GBM with an initial price of 10, $\sigma = 0.2$, 0.3, 0.4 and 0.5 and $\mu = 0.1$. A seed was used in order the pseudorandom generator to produce the same random variables, ϵ, for all four series

series, each time with a different w. In other words the identification mechanism performed a half million iterations.

When the HS pattern recognition mechanism was implemented on this simulated dataset, the pattern was on average identified in its normal and inverse form 21.77 % and 22.94 % of the time respectively! This is a very large figure, especially when it comes from randomly generated price series. This finding implies that the HS technical pattern can come into existence out of practically pure randomness. This also means that since the pattern can be identified in a random series it is reasonable to expect that the same stands for real series too. A similar report can be traced back in 1959 (Roberts 1959). Particularly Roberts mentions:

> ...the usual method of graphing stock prices gives a picture of successive levels rather than of changes, and levels can give an artificial appearance of pattern or trend. A second is that chance behavior itself produces patterns that invite spurious interpretations.

Under the concept of behavioral finance, bibliography provides evidence that investors can suffer from many cognitive biases resulting in irrational trading activity (Warneryd 2001). For example clustering illusion is a cognitive bias that makes man believe that he can observe patterns in a chaotic system. It seems that true chance looks too ordered to us and the title of the second chapter in Gilovich (1993) is characteristic: "*Something Out of Nothing: The Misperception and Misinterpretation of Random Data*". Additionally, the following excerpt from Zielonka (2004) is along the lines of our findings:

> ... even if stock prices completely followed a random walk, people would be able to convince themselves that there are patterns having predictive value. In laboratory experiments, subjects are reported to have found patterns in purely random sequences of stock prices.

Related is a statement from De Bondt (1998):

> Many people: 1. discover naïve patterns in past price movements, 2. share popular models of value, 3. are not properly diversified and 4. trade in suboptimal ways.

Since the HS pattern was successfully identified in randomly simulated price series it is worth to see how values of μ, σ and w affect the number of patterns identified. This relation is presented in Fig. **5.6**. Specifically, the number of identified HS tops and bottoms seem to be positively related with σ and negatively related with w. Increasing the volatility rate of a series, *ceteris paribus*, seems to increase the probability of forming regional locals that satisfy the aforesaid required criteria for HS recognition. Additionally, narrowing the size of the rolling window, results in more identified sequences of regional locals and thus greater probability of observing a pattern. The drift rate, μ, seems to affect differently the number of identified HS tops and bottoms patterns. Particularly, more HS tops patterns are traced as μ approaches zero. When its value diverts from zero, in either direction the number of recognized patterns reduces. Similar is the picture in HS bottom, with the difference that the number of identified patterns reduces less when μ increases.

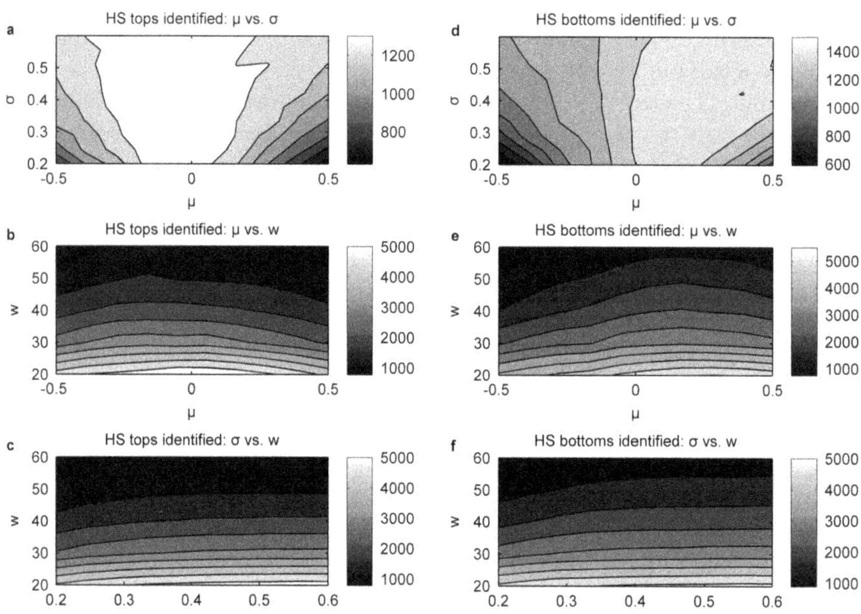

Fig. 5.6 Sensitivity of the number of identified HS tops and bottoms in respect to different values of μ, σ and w. (**a**) HS tops identified: μ vs. σ, (**b**) HS tops identified: μ vs. w, (**c**) HS tops identified: σ vs. w, (**d**) HS bottoms identified: μ vs. σ, (**e**) HS bottoms identified: μ vs. w, (**f**) HS bottoms identified: σ vs. w

5.3 Identifying the Double/Triple Tops/Bottoms

Triple tops (TT) and triple bottoms (TB) is another celebrated pair of zigzag technical patterns. Their description is very similar with that of HS tops and bottoms patterns. The crucial difference is that the three peaks (bottoms) in the TT (TB) are at the exact same level. Practically it is very difficult to find patterns that exhibit three peaks at the same price level, so the subsequent proposed identification mechanism is designed to locate sequences of three locals of the same type, that are within reasonable proximity. This is not the case for the two intervening bottoms (peaks) that can be observed at different levels. Since both HS and TT/TB signify the same future price evolution, it is of little importance whether a pattern is categorised to either type. Murphy (1986) characterizes TT/TB as a variation of HS pattern and comments on this similarity:

> Chartists often disagree as to whether a reversal pattern is a head and shoulders or a triple top. The argument is academic, because both patterns imply the exact same thing.

The notation used for the following conditions refers to the case of the TT (Fig. **5.7**) and TB (Fig. **5.8**) reversal patterns:

Condition 1. Trend Preexistence: The pattern is preceded by a generally positive or negative underlying trend. Equation (5.15) corresponds to TT and (5.16) to TB.

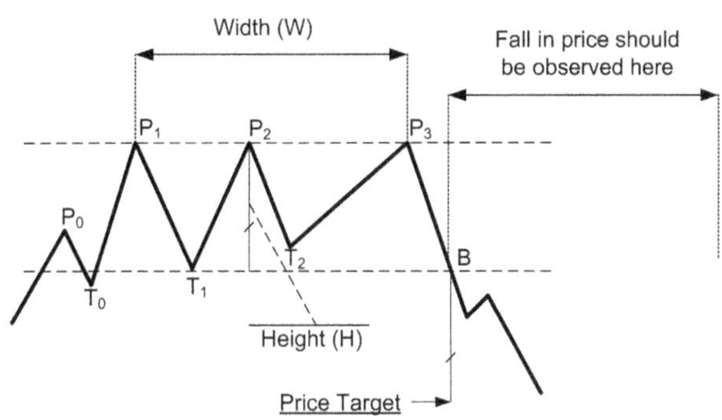

Fig. 5.7 Simplified illustration of the TT pattern

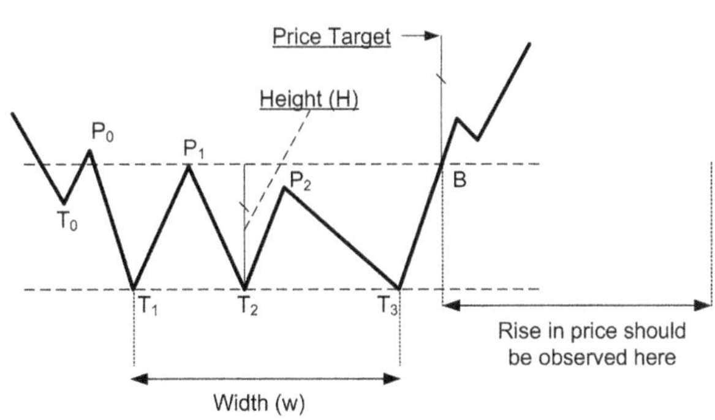

Fig. 5.8 Simplified illustration of the TB pattern

$$\text{TT}: \text{ uptrend if } P_1 > P_0 \ \& \ T_1 > T_0 \tag{5.15}$$

$$\text{TB}: \text{ downtrend if } P_1 < P_0 \ \& \ T_1 < T_0 \tag{5.16}$$

Condition 2. Balance: Three peaks (bottoms) are about the same level. In addition the second peak (bottom) should not be the maximum (minimum) local of the pattern. Otherwise it is a HS.

$$\text{TT}: \frac{\max(P_1,P_3) - \min(P_1,P_3)}{\min(P_1,P_3)} \leq 0.04 \ \& \ \text{P}_2 \leq \max(P_1,P_3) \tag{5.17}$$

$$\text{TB}: \frac{\max(T_1,T_3) - \min(T_1,T_3)}{\min(T_1,T_3)} \leq 0.04 \ \& \ T_2 \geq \min(T_1,T_3) \tag{5.18}$$

Condition 3. Intervening locals: In case of TT, the second trough must be at least at the same level with the first one and as close as possible. Otherwise there will be a penetration of the support level that the first trough indicates, and there will be a DB (and not TT) formation. Analogous, is the treatment in TB case.

$$T_2 \geq T_1 \tag{5.19}$$

$$(T_2 - T_1)/T_1 \leq 0.04 \tag{5.20}$$

This gives

$$T_1 \leq T_2 \leq 1.04T_1, \tag{5.21}$$

for the TT, whereas the corresponding condition for the TB is

$$P_2 \leq P_1 \leq 1.04P_2. \tag{5.22}$$

Condition 4. Symmetry: The time between the first two peaks (troughs) must not be more than 2.5 times the time between last two peaks (troughs) and vice versa.

$$\text{TT}: t_{P_2} - t_{P_1} < 2.5(t_{P_3} - t_{P_2}) \ \& \ t_{P_3} - t_{P_2} < 2.5(t_{P_2} - t_{P_1}) \tag{5.23}$$

$$\text{TB}: t_{T_2} - t_{T_1} < 2.5(t_{T_3} - t_{T_2}) \ \& \ t_{T_3} - t_{T_2} < 2.5(t_{T_2} - t_{T_1}) \tag{5.24}$$

Condition 5. Penetration: Let t_B denote the time at which the price falls below the support level indicated by points T_1 and T_2. This must not happen too long after the formation of the right peak (distance *w* in Fig. 5.7),

$$t_B < t_{P_3} + (t_{P_3} - t_{P_1}). \tag{5.25}$$

Similarly, the appropriate inequality for the TB is

$$t_B < t_{T_3} + (t_{T_3} - t_{T_1}). \tag{5.26}$$

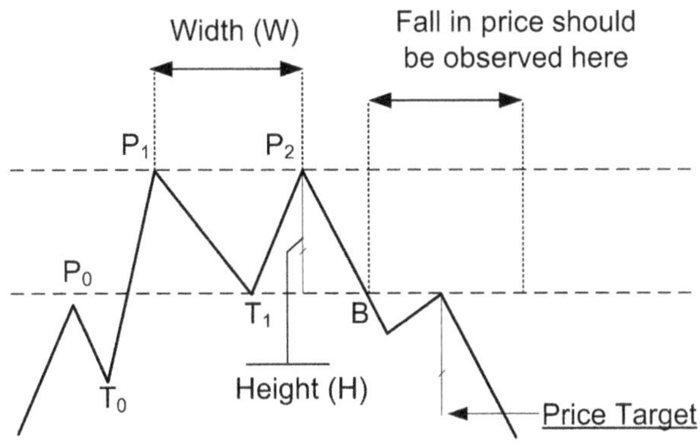

Fig. 5.9 Simplified illustration of the DT pattern

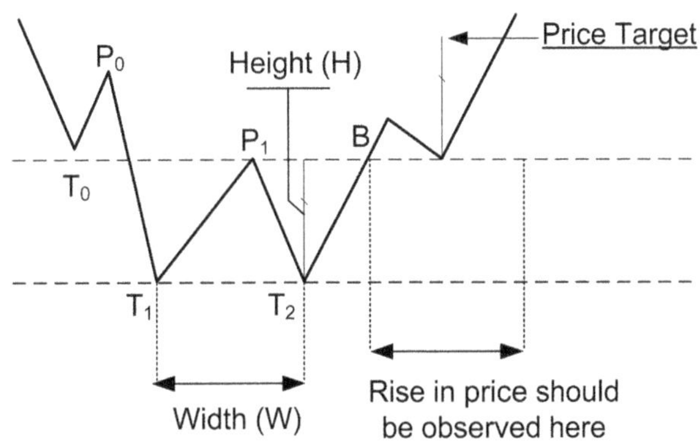

Fig. 5.10 Simplified illustration of the DB pattern

Bulkowski (2000) suggests setting the price target for the TT by subtracting the lowest low from the highest peak and subsequently subtracting this difference from the lowest low (Fig. 5.7). The reverse stands for the TB pattern (Fig. 5.8).

Similar to the aforesaid patterns, are the double tops (DT) and the double bottoms (DB) patterns. The resemblance with the letters "M" and "W" made these patterns known as an "M" and a "W" pattern respectively. At the case of the "M" pattern, price evolvement creates two subsequent peaks at about the same level with a local bottom between them. Later on, a beginning of a new downtrend takes place, and the "M" pattern signals a short position at the price level signified by the intervening trough. The DB pattern is a mirror image of the DT. Figures 5.9 and 5.10 present the characteristic components of the DT and DB patterns respectively. Similarly, the conditions adopted for their identification are the following:

Condition 1. Trend Preexistence: Same with the first condition applied in TT/TB [see (5.15) and (5.16)].

Condition 2. *Balance:* Price variation between two peaks should be less than 4%. Two peaks (bottoms) are about the same level.[8]

[8] Bulkowski (2000) suggests maximum price variations of 3% and 4% for DT and DB respectively.

$$\text{DT}: \ |P_1 - P_2|/\min(P_1, P_2) \leq 0.04 \tag{5.27}$$

$$\text{DB}: \ |T_1 - T_2|/\min(T_1, T_2) \leq 0.04 \tag{5.28}$$

Condition 3. Depth: There should be at least a 10 % decline (rise) between the two tops (bottoms) (Bulkowski 2000).

$$\text{DT}: (T_1 - P_1)/P_1 \leq -0.1 \tag{5.29}$$

$$\text{DB}: (P_1 - T_1)/T_1 \geq 0.1 \tag{5.30}$$

Condition 4. Penetration: For the DT (DB) case, let t_B denote the time at which the price falls below (rises above) the support (resistance) level indicated by point T_1 (P_1). This must not happen too long after the formation of the right peak (distance w in Figs. 5.9 and 5.10).

$$\text{DT}: \ t_B < t_{P_2} + (t_{P_2} - t_{P_1}) \tag{5.31}$$

$$\text{DB}: \ t_B < t_{T_2} + (t_{T_2} - t_{T_1}) \tag{5.32}$$

5.4 Identifying Flags, Pennants and Wedges

Flags, pennants and wedges are technical zigzag patterns whose formations require shorter time intervals compared with the other technical patterns examined previously. These patterns are considered as brief pauses in a dynamic market move (Murphy 1986). Particularly, flags and pennants appear at the midpoint of this dynamic trend, thus in technical jargon they are said to fly at half-mast (Bulkowski 2000; Pring 2002).When the market move is "bullish" then these pauses are reflections of controlled profit taking periods. Contrary, when the market move is "bearish" they are reflections of controlled digestion of losses (Pring 2002). Due to this, they are generally considered as trend-continuation patterns.

Flags and wedges realize a slope bias against the prevailing trend. This characteristic is rarer in the case of pennants, since they are generally characterized to be horizontal. In addition, Edwards and Magee (1997) mention that the slope is actually slightly in the same direction as the preceding trend instead of being against it. For a successful Flag formation, price should fluctuate within two parallel trendlines (Fig. 5.11). A pennant is identified when these trendlines converge and it is considered to be more horizontal (Fig. 5.12). The trendlines converge also in the case of a Wedge, but the pattern's specific characteristic is its apparent slope (Fig. **5.13**). More concretely, the two trendlines slope to the same direction (either

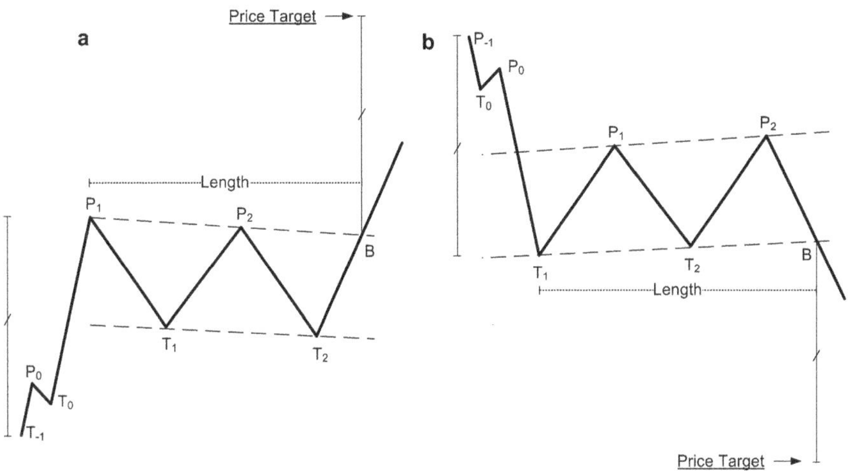

Fig. 5.11 Simplified illustrations of (**a**) descending (bullish) and (**b**) ascending (bearish) flags

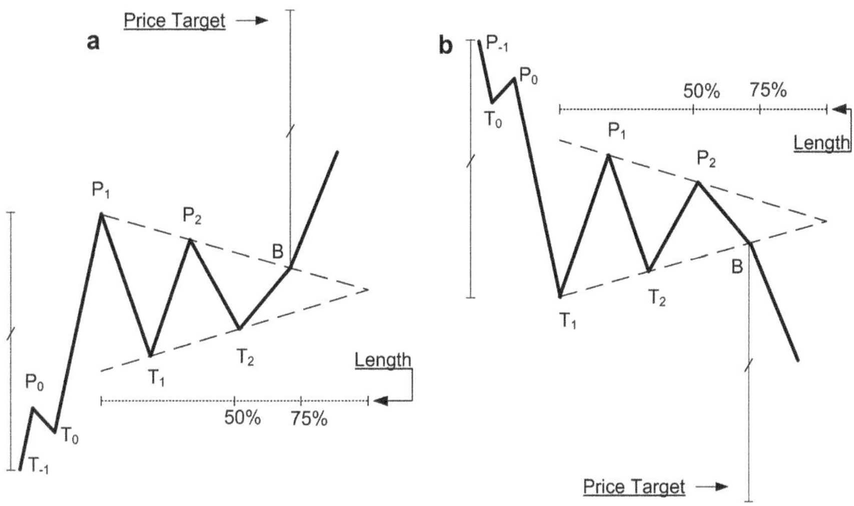

Fig. 5.12 Simplified illustrations of (**a**) bullish and (**b**) bearish pennants

both upwards or both downwards). Flags and pennants generally require up to 4 weeks (20 trading weeks) to form whereas wedges 3 weeks to 4 months.[9]

[9] The proposed maximum time required for a flag's formation varies, but not significantly, in bibliography. For example, regarding flag formations, Edwards and Magee (1997) and Bulkowski (2000) report a maximum duration of 3 weeks whereas Pring (2002) 3–5 weeks. Similar, divergences are also found in other aspects of identification guidelines provided in bibliography. In such cases we use average values in our methodology.

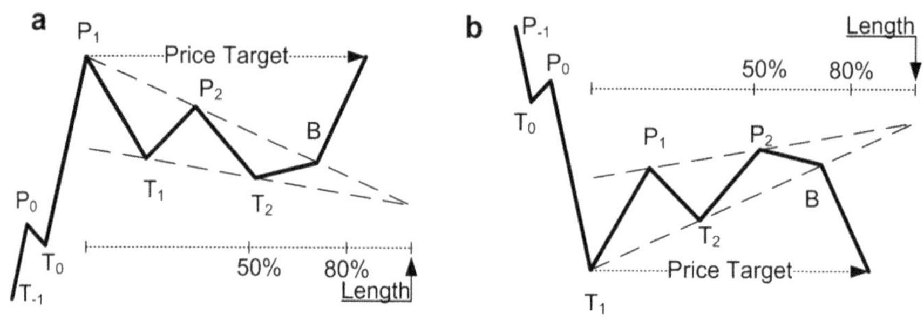

Fig. 5.13 Simplified illustrations of (**a**) falling (bullish) and (**b**) rising (bearish) wedges

The identification of flags, pennants and wedges is accomplished with processes based on the same notion with methods presented in Sects. 5.2 and 5.3. Again, $RW(\cdot)$ identifies regional locals. The difference here is that these patterns represent brief pauses in a dynamic market move. This means that regional locals inside the pattern should be identified with a short term rolling window, whereas locals before the patterns fully formation occur in longer time intervals. We deal with this problem by combining $RW(\cdot)$ and $PIPs(\cdot)$. Particularly, a short term rolling window is applied in $RW(\cdot)$ to identify sequences of four alternating regional locals. For a time period before the first local we adopt the $PIPs(\cdot)$ to spot the preceding three regional locals. Recall from the descriptions of these patterns that each pattern can be spotted either with an ascending or a descending form. The identification of each pattern is utilized when sequences of these regional locals satisfy a set of conditions. For the following proposed conditions, t^* is the time that the two trendlines intersect, B is the price at which a trading signal occurs and t_i is the time when the local i arises.

Subsequently, the conditions adopted for the identification of the two types of Flags are presented. For each set of equations the first corresponds to the descending (or bullish) flags and the second to the ascending (bearish) flags. These conditions are included in the $FLAGS(\cdot)$ (see Appendix 2):

Condition 1. Trend Preexistence: (5.33) corresponds to the bullish whereas (5.34) to the bearish flag.

$$\text{Flag}_{\text{Bull}} : P_1 > \max(P_0, T_0) \tag{5.33}$$

$$\text{Flag}_{\text{Bear}} : T_1 < \min(T_0, P_0) \tag{5.34}$$

Condition 2. P_0 (T_0) should be below (above) the lower (upper) line signaling the bullish (bearish) flag.

$$\text{Flag}_{\text{Bull}} : \max(P_0, T_0) < \frac{T_2 - T_1}{t_{T_2} - t_{T_1}}\left(t_{\max(P_0,T_0)} - t_{T_1}\right) + T_1 \tag{5.35}$$

$$\text{Flag}_{\text{Bear}} : \min(T_0, P_0) > \frac{P_2 - P_1}{t_{P_2} - t_{P_1}} \left(t_{\min(T_0, P_0)} - t_{P_1} \right) + P_1 \tag{5.36}$$

Condition 3. Slope of the two parallel lines is against the prevailing trend.

$$\text{Flag}_{\text{Bull}} : P_2 < P_1 \;\&\; T_2 < T_1 \tag{5.37}$$

$$\text{Flag}_{\text{Bear}} : P_2 > P_1 \;\&\; T_2 > T_1 \tag{5.38}$$

Condition 4. Parallel lines: These inequalities are set in order to define whether the two trendlines are almost parallel. In (5.39) and (5.40) t^* is the time (if any) the two trendlines intersect. Three different scenarios are possible. First, in the ideal case that the two lines have (almost) the same slope t^* is not defined ($t^* = \infty$). In other two scenarios where the two lines converge either in the future or in the past, the inequality ensures that this occurs significantly later (before) the pattern's formation.

$$\text{Flag}_{\text{Bull}} : \left| t^* - (t_{P_1} + t_{T_2})/2 \right| > 5(t_{T_2} - t_{P_1}) \tag{5.39}$$

$$\text{Flag}_{\text{Bear}} : \left| t^* - (t_{T_1} + t_{P_2})/2 \right| > 5(t_{P_2} - t_{T_1}) \tag{5.40}$$

Condition 5. Maximum length: As we have already mentioned flags are short term patterns and therefore the following conditions regarding their maximum length are adopted. Bulkowski (2000) suggests a maximum length of 3 weeks whereas Pring (2002) reports that flags' maximum length can be between 3 and 5 weeks. We set the maximum length criterion to 4 trading weeks (20 trading days) for both bullish (5.41) and bearish (5.42) flags.

$$\text{Flag}_{\text{Bull}} : L = t_B - t_{P_1} \leq 20 \tag{5.41}$$

$$\text{Flag}_{\text{Bear}} : L = t_B - t_{T_1} \leq 20 \tag{5.42}$$

Condition 6. Penetration: Upward and downward penetration for descending and ascending flags respectively must occur no later than

$$\text{Flag}_{\text{Bull}} : t_B < t_{T_2} + (t_{T_2} - t_{P_1}) \tag{5.43}$$

and

$$\text{Flag}_{\text{Bear}} : t_B < t_{P_2} + (t_{P_2} - t_{T_1}). \tag{5.44}$$

Condition 7. Since these patterns represent brief pauses in a dynamic market move the length of the preceding trend should be larger than the length of the formation.

$$\text{Flag}_{\text{Bull}} : t_{P_1} - t_{T_{-1}} > 2(t_B - t_{P_1}) \tag{5.45}$$

$$\text{Flag}_{\text{Bear}} : t_{T_1} - t_{P_{-1}} > 2(t_B - t_{T_1}) \tag{5.46}$$

Again, (5.45) correspond to the bullish and (5.46) to the bearish flags.

Regarding pennants and wedges most criteria are similar with those presented for the flags pattern. Cralification is needed for the "penetration" criterion of both patterns. Concerning pennants, this criterio is needed in order to make sure that price penetrates the two trendlines, and thus generates a signal, after the midpoint and before the 75 % of the pattern's length (Fig. 5.12). For wedges the penetration must occur no later than the 80 % of the pattern's length (Bulkowski 2000) (Fig. **5.13**).

In the same spirit, the conditions we adopt for the identifications of pennants are the following:

Condition 1. Trend Preexistence: P_0 should be below the lower local of the bullish pennant (*5.47*). Analogously, (5.48) corresponds to the bearish pennants.

$$\text{Pennant}_{\text{Bull}} : \min(P_1, T_1) > \max(P_0, T_0) \tag{5.47}$$

$$\text{Pennant}_{\text{Bear}} : \max(T_1, P_1) < \min(T_0, P_0) \tag{5.48}$$

Condition 2. Trendlines must converge: The same condition (5.49) for both types of pennants must hold.

$$P_2 < P_1 \; \& \; T_2 > T_1 \tag{5.49}$$

In addition the slope is actually slightly in the same direction as the preceding trend instead of against it. For bullish pennants this is

$$\text{Pennant}_{\text{Bull}} : \frac{T_2 - T_1}{t_{T_2} - t_{T_1}} \geq \left| \frac{P_2 - P_1}{t_{P_2} - t_{P_1}} \right|, \tag{5.50}$$

Whereas for bearish pennants

$$\text{Pennant}_{\text{Bear}} : \frac{T_2 - T_1}{t_{T_2} - t_{T_1}} \leq \left| \frac{P_2 - P_1}{t_{P_2} - t_{P_1}} \right|. \tag{5.51}$$

Additionally, trendlines should intersect soon enough, say

$$\text{Pennant}_{\text{Bull}} : t^* - t_{P_1} < 2(t_{T_2} - t_{P_1}) \Leftrightarrow t^* < 2t_{T_2} - t_{P_1} \tag{5.52}$$

$$\text{Pennant}_{\text{Bear}} : t^* - t_{T_1} < 2(t_{P_2} - t_{T_1}) \Leftrightarrow t^* < 2t_{P_2} - t_{T_1} \tag{5.53}$$

Condition 3: Maximum length (same with flags)*:*

$$\text{Pennant}_{\text{Bull}} : L = t_B - t_{P_1} \leq 20 \tag{5.54}$$

$$\text{Pennant}_{\text{Bearl}} : L = t_B - t_{T_1} \leq 20 \tag{5.55}$$

Condition 4: Penetration: We have already described that penetration should occur between 50% and 75% of the pattern's length.

$$\text{Pennant}_{\text{Bull}} : 0.5\left(t^* - t_{P_1}\right) < t_B - t_{P_1} < 0.75\left(t^* - t_{P_1}\right) \tag{5.56}$$

$$\text{Pennant}_{\text{Bear}} : 0.5\left(t^* - t_{T_1}\right) < t_B - t_{T_1} < 0.75\left(t^* - t_{T_1}\right) \tag{5.57}$$

Condition 5: Since these patterns represent brief pauses in a dynamic market move the length of the preceding trend should be larger than the length of the formation.

$$\text{Pennant}_{\text{Bull}} : t_{P_1} - t_{T_{-1}} > 2(t_B - t_{P_1}) \tag{5.58}$$

$$\text{Pennant}_{\text{Bear}} : t_{T_1} - t_{P_{-1}} > 2(t_B - t_{T_1}) \tag{5.59}$$

Simplified illustrations of the two wedges versions are presented in Fig. 5.13. Conditions adopted for the identification of wedges are presented subsequently. For each pair of the following inequalities the first inequality corresponds to the bullish whereas the second to the bearish wedge.

Condition 1. Trend Preexistence: P_0 (T_0) should be below (above) the lower (higher) local of the bullish (bearish) wedge.

$$\text{Wedge}_{\text{Bull}} : \text{T}_2 > \max(P_0, T_0) \tag{5.60}$$

$$\text{Wedge}_{\text{Bear}} : \text{P}_2 < \min(T_0, P_0) \tag{5.61}$$

Condition 2. Down-sloping (up-sloping) trendlines must converge

$$\text{Wedge}_{\text{Bull}} : 0 > \frac{T_2 - T_1}{t_{T_2} - t_{T_1}} > \frac{P_2 - P_1}{t_{P_2} - t_{P_1}} \tag{5.62}$$

$$\text{Wedge}_{\text{Bear}} : \frac{T_2 - T_1}{t_{T_2} - t_{T_1}} > \frac{P_2 - P_1}{t_{P_2} - t_{P_1}} > 0 \tag{5.63}$$

Additionally, trendlines should intersect soon enough (same with pennants), say

$$\text{Wedge}_{\text{Bull}} : t^* - t_{P_1} < 2(t_{T_2} - t_{P_1}) \Leftrightarrow t^* < 2t_{T_2} - t_{P_1} \tag{5.64}$$

$$\text{Wedge}_{\text{Bear}} : t^* - t_{T_1} < 2(t_{P_2} - t_{T_1}) \Leftrightarrow t^* < 2t_{P_2} - t_{T_1} \tag{5.65}$$

Condition 3. Minimum and maximum duration:

$$\text{Wedge}_{\text{Bull}} : 20 \leq t_B - t_{P_1} \leq 80 \tag{5.66}$$

$$\text{Wedge}_{\text{Bear}} : 20 \leq t_B - t_{\text{T}_1} \leq 80 \tag{5.67}$$

Condition 4. Penetration: The most powerful breakouts occur between 50 % and 80 % of the distance to the apex (centered around 65 %) (Bulkowski 2000).

$$\text{Wedge}_{\text{Bull}} : 0.5\left(t^{*} - t_{P_1}\right) < t_B - t_{P_1} < 0.8\left(t^{*} - t_{P_1}\right) \tag{5.68}$$

$$\text{Wedge}_{\text{Bear}} : 0.5\left(t^{*} - t_{T_1}\right) < t_B - t_{T_1} < 0.8\left(t^{*} - t_{T_1}\right) \tag{5.69}$$

5.5 Choice of *w*

The choice of w used in $RW(\cdot)$ is crucial in the patterns' identification process. For example, if a relative large value is used for tracing a short-term pattern (e.g. flags or pennants) the algorithm will return sequences of alternating regional locals with distances that do not satisfy the aforementioned conditions. We choose the value of w based again on assertions found in technical manuals which imply the average distance between two successive locals. For instance, Bulkowski (2000) provides such statistics for the zigzag patterns examined in this chapter.

For example, many technicians consider a month (20 trading days) to be the minimum distance between the two bottoms in DB. In addition, best gains come from formations with bottoms about 3 months apart on average, whereas the average formation length is 2 months (Bulkowski 2000). The last statement implies that DB realize two regional troughs (T_1 and T_2 in Fig. 5.10) with a time interval of 40 trading days. Considering that within this time period peak P_1 also occurs, the average distance between two subsequent regional locals (regardless their type) is 20 trading days. Table 5.2 presents the average formations' lengths $\bar{l}$ reported in bibliography for HS, TT/TB, DT/DB, flags, pennants and wedges. Third column illustrates the number of regional locals n appearing within these time spans. Under this concept, the implied average distance between two successive locals is $l^{*} = \bar{l}/(n-1)$. Last column presents the value of the input variable w we are

Table 5.2 The choice of w for the identification of each zigzag pattern

Pattern	$\bar{l}$	n	$\bar{l}/(n-1) = l^{*}$	w
HS	2 months	5	40/4 = 10	7
TT/TB	4 months	5	80/4 = 20	15
DT/DB	2 months	3	40/2 = 20	15
Flags	11 days	4	11/3 = 3.67	3
Pennants	10 days	4	10/3 = 3.33	2
Wedges	1.5 month	4	30/3 = 10	7

$\bar{l}$ patterns' average duration, n number of regional locals within $\bar{l}$, l^{*} implied average distance between two successive locals, w input variable in $RW(\cdot)$

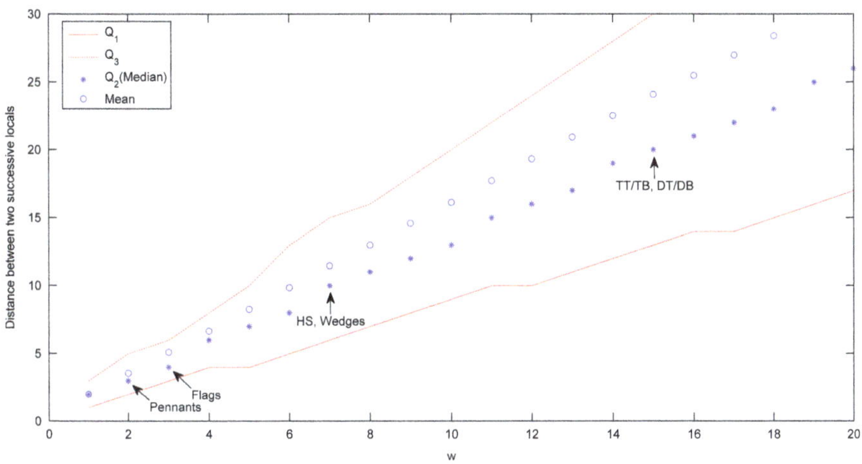

Fig. 5.14 Q_1, Q_2 and Q_3 are the first, second and third quartile respectively. *Arrows* show the suggested value of w for the identification of the examined zigzag patterns

about to use in $RW(\cdot)$ for identifying the zigzag patterns presented in this chapter. These values result by the following simulation.

For 100 combinations of different drift (μ) and volatility (σ) rates, 100 simulated price paths of 1000 observations with the discrete version of GBM (5.14) were generated. On this simulated dataset, the $RW(\cdot)$ with different values of w was adopted and some distributional characteristics (quartiles and mean) of the differences between two successive regional locals were calculated (Fig. 5.14). These distributions are unimodal characterized by positive skew with significant presence of outliers in right tails. However, for large datasets, in presence of outliers the median is robust, while the mean is rather sensitive. For every zigzag pattern, the choice of w, resulted by finding its value which returns a median closer to l^*. Our intention is to use the w which will increase the chances of observing a pattern with an average duration $\bar{l}$. Alternatively, we can perform the same process on the examined real price series and choose the corresponding value for w.

5.6 Design of Trading Rules

After the development of a robust pattern identification mechanism, the design of the trading rules is of prime importance. For obtaining persuasive results, these rules must be set with precise criteria in a manner that simulates the trading activities of an investor in real life. Empirical results presented in this book are based on two different categories of trading strategies. In the simplest category, after the penetration of the neckline, the corresponding investment position is taken

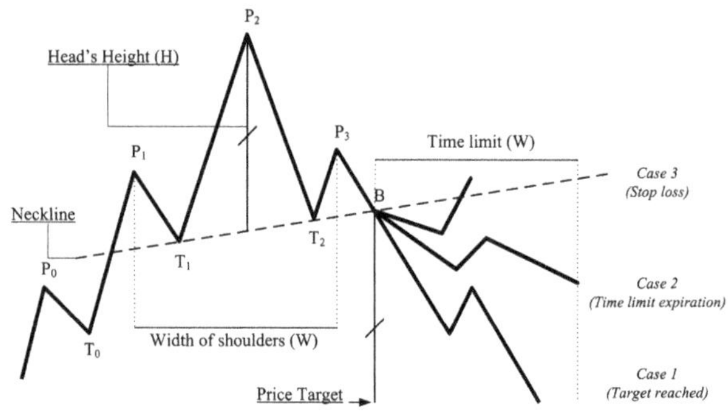

Fig. 5.15 Trading rule design for the HS tops pattern. This figure illustrates all three possible alternative price evolutions after the pattern's confirmation

and held for a predefined period.[10] In addition a more case sensitive trading rule is developed to capture the investment behaviour of a technician in a more realistic manner. The description of this trading rule is presented in the following analysis.

In case of HS tops (HS bottoms) the investor takes a short (long) position as soon as price breaches the neckline downwards (upwards). The initial position is closed (reversed) at a time determined by the subsequent price behavior. In practice, after the pattern's confirmation, the price path can evolve according to one of the three, following, alternative ways (Fig. 5.15):

- Case 1: The price evolves exactly as TA asserts. In this case, the initial position is closed as soon as the drop (rise) in the price exceeds the target set by the HS tops (HS bottoms) pattern.
- Case 2: After the penetration, the price stays below the neckline but does not reach the price target as asserted by TA. In such case, the initial short position is closed as soon as the time limit set by the shoulders' width (W) expires.
- Case 3: After the pattern's confirmation, the price increases above the limit imposed by the neckline, at least for n consecutive days. In this case the short position is closed after the nth day or at the moment that the price increase results in a predetermined accepted *sl%* loss compared with the price the investor initially shorted.

For brevity, the above description and (Fig. 5.15) refer only to the HS tops pattern. For the inverse HS form the corresponding cases can be also defined analogously.

The design of trading rules for the rest zigzag patterns is based on the same notion. Particularly, for each pattern, i, the *price target* (PT_i), the *time limit* (TL_i) and a stop loss condition must be specified. These values result again from conventions found in technical manuals. Regarding TT and TB (Figs. 5.7 and 5.8), the time limit is set by adding the pattern's width ($W_{TT} = t_{P_3} - t_{P_1}$ and $W_{TB} = t_{T_3} - t_{T_1}$) to the time the pattern is confirmed, t_B. Price target is defined by the pattern's height ($PT_{TT} = B - H_{TT}$ and $PT_{TB} = B + H_{TB}$). Table **5.3** presents the

[10] This predefined period could be arbitrarily chosen like the holding periods used in HSAR (Chap. 4) or it could be equal with the patterns width.

Table 5.3 Defining the price target and time limit for each version of each zigzag pattern

Pattern (i)	H_i	W_i	PT_i	TL_i
HS_{Tops}	$P_2 - \frac{T_2-T_1}{t_{T_2}-t_{T_1}}(t_{P_2} - t_{T_1}) + T_1$	$t_{P_3} - t_{P_1}$	$B - H_{HS}$	$t_B + W_{HS}$
$HS_{Bottoms}$	$\frac{P_2-P_1}{t_{P_2}-t_{P_1}}(t_{T_2} - t_{P_1}) + P_1 - T_2$	$t_{T_3} - t_{T_1}$	$B + H_{HS}$	$t_B + W_{HS}$
TT	$\max(P_1, P_3) - T_1$	$t_{P_3} - t_{P_1}$	$T_1 - H_{TT}$	$t_B + W_{TT}$
TB	$P_1 - \min(T_1, T_3)$	$t_{T_3} - t_{T_1}$	$P_1 + H_{TB}$	$t_B + W_{TB}$
DT	$\max(P_1, P_2) - T_1$	$t_{P_2} - t_{P_1}$	$T_1 - H_{DT}$	$t_B + W_{DT}$
DB	$P_1 - \min(T_1, T_2)$	$t_{T_2} - t_{T_1}$	$P_1 + H_{DB}$	$t_B + W_{DB}$
$Flag_{Bull}$			$B + P_1 - T_{-1}$	$t_B + t_{P_1} - t_{T_{-1}}$
$Flag_{Bear}$			$B - (P_{-1} - T_1)$	$t_B + t_{T_1} - t_{P_{-1}}$
$Pennant_{Bull}$			$B + P_1 - T_{-1}$	$t_B + t_{P_1} - t_{T_{-1}}$
$Pennant_{Bear}$			$B - (P_{-1} - T_1)$	$t_B + t_{T_1} - t_{P_{-1}}$
$Wedge_{Bull}$			P_1	$2t_B - t_{P_1}$
$Wedge_{Bear}$			T_1	$2t_B - t_{T_1}$

i pattern's name, H_i *i*th pattern's height, W_i *i*th pattern's width, PT_i *i*th pattern's price target, TL_i *i*th pattern's time limit

price targets (PT_i) and the time limits (TL_i) for each pattern (i) examined along with the corresponding heights and widths wherever applicable.

Having defined the trading rules, the subsequent section examines the price behaviour after the identification of each type of the examined patterns.

5.7 Assessing the Predictive Performance

Let t_{sl} and t_n be the two thresholds presented previously in Sect. 5.5 that define the day the initial position is closed after the identification of a zigzag pattern. For different combinations of threshold values $t_{sl} \in \{-0.15 : 0.01 : 0\}$, and $t_n \in \{1 : 1 : 10\}$, Figs. **5.16**a, d present the relative frequencies of case 1, Figs. **5.16**b, e present the relative frequencies of cases 1 and 2 jointly, and Figs. **5.16**c, f the average trading returns of all cases for NASDAQ and NYSE components. Price behaviour under case 1 is presented because it represents the proportion of cases where price evolves as TA asserts. In addition, cases 1 and 2 are jointly presented, because in these cases it is expected to observe price paths which will return positive trading returns, i.e. the price target is reached within the time limit (case 1), or the position is closed without the price meeting the price target nor a stop loss condition is being activated within the time limit. Finally, average trading returns of all cases are being presented to observe the overall performance of each examined pattern.

Figures 5.16a, d imply that the relative frequencies of case 1 can be increased by increasing the values of t_n and t_{sl} (in absolute terms) thresholds. In particular, conditioning a stop loss threshold, increasing the value of t_n, increases the proportions of case 1. This positive relation is more apparent when greater in absolute

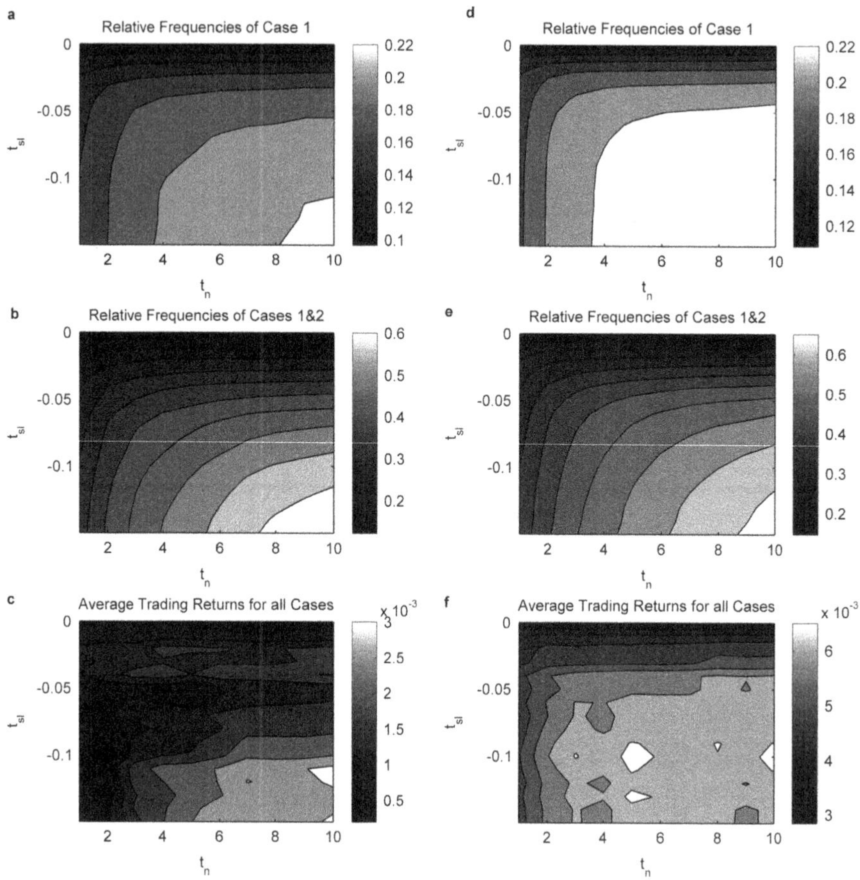

Fig. 5.16 Sensitivity of the relative frequencies of cases 1, cases 1 and 2 jointly and average trading returns of all cases for different parameters of stop loss conditions. (**a**, **b**, **c**) NASDAQ components, (**d**, **e**, **f**) NYSE components

terms t_{sl} values are used. This behaviour also appears when, conditioning the t_n threshold, the t_{sl} in absolute terms is increased. Again, this positive relation is stronger when greater t_n values are used. These results imply that allowing more flexibility, by using greater threshold values, in price evolution after a zigzag pattern's fully formation, increases the chances of observing subsequent price behaviour according to case 1. As illustrated, these relative frequencies are greater than 20 % in some stop-loss parameters combinations, which means that more than one out of five times where zigzag patterns identified, present subsequent price behaviour as asserted by TA. Figures 5.16b, e exhibit also consistency for both datasets used in terms of how relative frequencies of cases 1 and 2 jointly behave to different threshold values. This price behaviour is similar to the one presented in Figs. 5.16a, d. As expected, average trading returns exhibit a similar relation to the adopted threshold values, although this relation is not as smooth as in the case of frequencies, with the average trading returns observed in NYSE dataset being generally superior to those resulted by using the NASDAQ dataset (Figs. 5.16c, f). However, recall that these returns represent the average returns generated by identified patterns and they are not expressed on a daily basis. Bearing also in mind the fact that no transaction costs were considered, these returns, although positive, they might not be sufficient large to constitute zigzag patterns efficient.

Tables 5.4 and 5.5 scrutiny further the patterns' performance on both datasets for two particular threshold values, $t_n = 2$ and $t_{sl} = -0.04$. In addition, for each

Table 5.4 Zigzag pattern's results from NASDAQ components and simulated series. $t_n = 2$ and $t_{sl} = -0.04$

Case (i)	$f_i(\%)$	$r_i(\%)$	$f_i(\%)$	$r_i(\%)$
Panel A: HS	HS_{Tops} : $n_p = 908$, $n_s = 213$		$HS_{Bottoms}$: $n_p = 947$, $n_s = 215$	
1	18.17 (13.75)	14.41 (17.86)	19.96 (24.88)	11.41 (15.41)
2	14.87 (15.58)	5.99 (10.18)	15.21 (8.70)	3.83 (5.89)
3	66.96 (70.67)	−4.51 (−5.00)	64.84 (66.42)	−4.55 (−5.15)
Total	100 (100)	0.50 (0.50)	100 (100)	−0.09 (0.93)
Panel B: TTTB	TT : $n_p = 19$, $n_s = 19$		TB : $n_p = 12$, $n_s = 11$	
1	21.05 (15.94)	15.33 (21.8)	8.33 (0.00)	19.01 (n/a)
2	15.79 (3.62)	10.37 (10.74)	0.00 (0.00)	n/a (n/a)
3	63.16 (80.43)	−5.14 (−5.82)	91.67 (100)	0.19 (−0.20)
Total	100 (100)	1.62 (−0.81)	100 (100)	1.75 (−0.20)
Panel C: DTDB	DT : $n_p = 205$, $n_s = 128$		DB : $n_p = 223$, $n_s = 141$	
1	15.61 (14.05)	18.15 (21.24)	1.35 (0.21)	4.23 (8.21)
2	11.22 (12.52)	8.34 (9.65)	0.00 (0.00)	n/a (n/a)
3	73.17 (73.44)	−6.18 (−5.86)	98.65 (99.79)	−0.10 (0.01)
Total	100 (100)	−0.75 (−0.11)	100 (100)	−0.04 (0.03)
Panel D: Flags	$Bear$: $n_p = 40$, $n_s = 37$		$Bull$: $n_p = 361$, $n_s = 166$	
1	15.00 (7.97)	12.04 (21.85)	13.30 (15.87)	19.79 (25.18)
2	7.50 (11.38)	11.03 (16.21)	13.30 (9.92)	7.97 (12.09)
3	77.50 (80.65)	−5.58 (−5.65)	73.41 (74.21)	−5.33 (−5.83)
Total	100 (100)	−1.69 (−0.97)	100 (100)	−0.22 (0.87)
Panel E: Pennants	$Bear$: $n_p = 43$, $n_s = 36$		$Bull$: $n_p = 58$, $n_s = 50$	
1	11.63 (3.25)	38.38 (27.12)	13.79 (12.58)	18.93 (25.43)
2	13.95 (17.21)	17.50 (20.20)	15.52 (11.04)	7.27 (9.67)
3	74.42 (79.54)	−5.71 (−5.40)	70.69 (76.38)	−4.15 (−5.35)
Total	100 (100)	2.65 (0.06)	100 (100)	0.80 (0.18)
Panel F: Wedges	$Bear$: $n_p = 25$, $n_s = 23$		$Bull$: $n_p = 27$, $n_s = 25$	
1	36.00 (30.62)	12.44 (15.88)	33.33 (32.10)	13.46 (13.11)
2	16.00 (6.46)	3.53 (7.87)	11.11 (6.53)	3.48 (8.90)
3	48.00 (62.92)	−5.33 (−5.55)	55.56 (61.36)	−4.73 (−5.44)
Total	100(100)	2.48 (1.87)	100 (100)	2.25 (1.45)

i case 1, 2 or 3 describing the price evolution after identification of each pattern, $f_i(\%)$ relative frequency of each case, $r_i(\%)$ average return for each case, n_p number of identified patterns in original series, n_s number of original price series where at least one pattern was identified. Numbers in parenthesis are the corresponding results from simulated series

Table 5.5 Zigzag pattern's results from NYSE components and simulated series

Case (i)	$f_i(\%)$	$r_i(\%)$	$f_i(\%)$	$r_i(\%)$
Panel A: HS	$HS_{Tops}: n_p = 2{,}151,\ n_s = 490$		$HS_{Bottoms}: n_p = 2{,}485,\ n_s = 499$	
1	20.41 (15.32)	10.58 (12.06)	23.70 (26.02)	8.19 (10.88)
2	16.36 (17.01)	3.90 (5.70)	20.72 (11.99)	3.54 (3.89)
3	63.23 (67.68)	−3.56 (−3.90)	55.57 (61.99)	−3.44 (−3.89)
Total	100 (100)	0.54 (0.18)	100 (100)	0.77 (0.89)
Panel B: TTTB	$TT: n_p = 74,\ n_s = 68$		$TB: n_p = 67,\ n_s = 64$	
1	24.32 (14.27)	13.10 (15.8)	2.99 (0.10)	7.55 (10.62)
2	13.51 (8.36)	4.89 (5.68)	0.00 (0.00)	n/a (n/a)
3	62.16 (77.37)	−4.79 (−5.20)	97.01 (99.9)	−0.13 (0.11)
Total	100 (100)	0.87 (−1.29)	100 (100)	0.10 (0.12)
Panel C: DTDB	$DT: n_p = 445,\ n_s = 286$		$DB: n_p = 467,\ n_s = 299$	
1	19.55 (13.22)	16.73 (17.53)	1.50 (0.12)	7.53 (11.63)
2	13.03 (16.58)	5.89 (6.56)	0.00 (0.02)	n/a (4.45)
3	67.42 (70.20)	−5.59 (−5.39)	98.50 (99.85)	−0.05 (0.01)
Total	100 (100)	0.27 (−0.38)	100 (100)	0.07 (0.02)
Panel D: Flags	$Bear: n_p = 49,\ n_s = 45$		$Bull: n_p = 867,\ n_s = 416$	
1	12.24 (9.70)	10.15 (14.76)	14.65 (16.10)	14.89 (18.65)
2	4.08 (13.22)	34.03 (10.51)	19.72 (14.34)	6.21 (9.01)
3	83.67 (77.07)	−4.67 (−4.54)	65.63 (69.56)	−4.54 (−4.78)
Total	100 (100)	−1.28 (−0.68)	100 (100)	0.43 (0.97)
Panel E: Pennants	$Bear: n_p = 75,\ n_s = 72$		$Bull: n_p = 157,\ n_s = 131$	
1	8.00 (4.47)	25.96 (18.09)	8.92 (14.78)	15.57 (19.02)
2	13.33 (17.90)	8.40 (14.47)	16.56 (13.83)	8.45 (8.56)
3	78.67 (77.63)	−4.34 (−4.26)	74.52 (71.39)	−3.68 (−4.24)
Total	100 (100)	−0.22 (0.09)	100 (100)	0.05 (0.97)
Panel F: Wedges	$Bear: n_p = 80,\ n_s = 77$		$Bull: n_p = 71,\ n_s = 65$	
1	26.25 (30.90)	14.82 (11.03)	40.85 (39.16)	7.73 (9.73)
2	11.25 (9.95)	4.20 (4.37)	11.27 (10.50)	1.50 (3.69)
3	62.50 (59.15)	−4.53 (−4.83)	47.89 (50.34)	−4.22 (−4.84)
Total	100 (100)	1.53 (0.98)	100 (100)	1.30 (1.76)

$t_n = 2$ and $t_{sl} = -0.04$.
As for Table **5.4**

original price series, 10 artificial series by using GBM (5.13) are generated. Simulated series have the same initial price, length, drift and volatility rate with the original series. The breakdown of frequencies and average trading returns to cases 1, 2 and 3 (Sect. 5.5) for bearish and bullish versions of the examined zigzag patterns are tabulated.

For example, Panel A in Table 5.4 focuses on the HS pattern. Particularly, the normal form of HS (HS_{Tops}) was identified 908 times in 213 out of 232 series. After the each pattern's confirmation, price evolved according to cases 1, 2 and 3, 18.17 %, 14.87 % and 66.96 % of time respectively. A common finding across all examined patterns is that proportions corresponding to case 3 are greater than those for cases 1 and 2, with the case of meeting the price target (case 1) being more frequent than case 2 in most patterns. This means that patterns do not perform as TA asserts most of the time, and generally $f_3 \gg f_1 > f_2$ holds. The same picture is repeated in simulated series. This similarity is being statistically assessed and discussed later in Table 5.7. However, when they do perform as expected, they generate positive average returns which are greater than the corresponding positive returns of cases 2 and of course greater than the usually negative returns of cases 3; i.e. $r_1 > r_2 > 0 > r_3$ usually holds. Returning in the case of HS_{Tops} $r_1 = 14.41\%$, $r_2 = 5.99\%$ and $r_3 = -4.51\%$. Although return results per case are rational, frequency results can raise an important issue in investors' decision making process. Reported results indicate that price evolution following patterns' confirmations does not behave as TA asserts most of the time. Investor lacking the required scientific background, or suffering from various cognitive biases may focus *ex-post* only on the small part of the whole, real picture; the part that is suitable to them which encloses the cases where patterns behave as TA asserts and are profitable.

Subsequently it is examined if the reported price behaviour after a pattern's confirmation is similar in real and simulated price series. This examination has been carried by adopting two statistical tests; x^2-test for homogeneity of proportions and z-test for comparing proportions. Table 5.6 presents the proportions of instances, f_i, where price behaviour following a pattern's formation is allocated to case i, $i \in \{1, 2, 3\}$, f_i corresponds to proportions obtained from the real price series whereas $\widetilde{f}_i$ to those obtained from simulated series. These proportions correspond to both datasets jointly, in order to increase the number of patterns identified, n_p, and thus ensure that conditions to implement the aforementioned statistical tests are fulfilled.[11]

Table 5.7 presents the results (chi-squared and p-values) from x^2-test for homogeneity of proportions. The null hypothesis ($H_0^{1,2,3}$) states that conditioning a pattern type and version,[12] the allocation of price behaviour to different cases in real and simulated datasets is homogenous (5.70).

$$H_0^{1,2,3}: f_1 = \widetilde{f}_1, \; f_2 = \widetilde{f}_2 \text{ and } f_3 = \widetilde{f}_3 \tag{5.70}$$

The alternative hypothesis ($H_a^{1,2,3}$) is that at least one of the equalities mentioned in the null hypothesis does not hold. This test is implemented 12 times (six types of patterns times two versions of each pattern). Results are not reported for TB and DB cases because conditions for performing the x^2-test are not satisfied. From the remaining 10 tests, 6 of them rejected the null for a significance level of 5 %. This implies that in 4 instances (almost half the time) price evolution after the fully formation of the pattern behaves similarly in real and simulated series.

---

[11] Especially for the z-test for comparing proportions n_p should be greater than 30 in order to apply the central limit theorem.

[12] For example, *HS* is a type of a pattern which has two versions HS_{Tops} and $HS_{Bottoms}$.

Table 5.6 Proportions of cases 1, 2 and 3 reported in real and simulated price series

Case (i)	f_i	$\widetilde{f}_i$	f_i	$\widetilde{f}_i$
Panel A: HS	HS_{Tops}		$HS_{Bottoms}$	
1	19.75 %	14.82 %	22.67 %	25.66 %
2	15.92 %	16.55 %	19.20 %	10.95 %
3	64.33 %	68.63 %	58.13 %	63.39 %
n_p	3,059	54,806	3,432	60,531
Panel B: TTTB	*TT*		*TB*	
1	23.66 %	14.60 %	3.80 %	0.08 %
2	13.98 %	7.44 %	0.00 %	0.00 %
3	62.37 %	77.96 %	96.20 %	99.92 %
n_p	93	1,425	79	1,242
Panel C: DTDB	*DT*		*DB*	
1	18.31 %	13.49 %	1.45 %	0.15 %
2	12.46 %	15.23 %	0.00 %	0.02 %
3	69.23 %	71.28 %	98.55 %	99.83 %
n_p	650	17,267	690	17,931
Panel D: Flags	*Bear*		*Bull*	
1	13.48 %	9.14 %	14.25 %	16.03 %
2	5.62 %	12.63 %	17.83 %	13.01 %
3	80.90 %	78.24 %	67.92 %	70.96 %
n_p	89	1,893	1,228	18,467
Panel E: Pennants	*Bear*		*Bull*	
1	9.32 %	4.05 %	10.23 %	14.12 %
2	13.56 %	17.66 %	16.28 %	12.99 %
3	77.12 %	78.29 %	73.49 %	72.89 %
n_p	118	2,741	215	3,017
Panel F: Wedges	*Bear*		*Bull*	
1	28.57 %	30.80 %	38.78 %	37.13 %
2	12.38 %	8.70 %	11.22 %	9.36 %
3	59.05 %	60.49 %	50.00 %	53.50 %
n_p	105	1,172	98	1,228

Table 5.7 x^2-test for homogeneity, 2 degrees of freedom

	Bearish		Bullish	
	x^2	p-Value	x^2	p-Value
HS	55.21	**0.0000**	219.88	**0.0000**
TTTB	12.25	**0.0022**	n/a	n/a
DTDB	14.15	**0.0008**	n/a	n/a
Flags	5.19	0.0748	23.96	**0.0000**
Pennants	8.48	**0.0144**	3.84	0.1467
Wedges	1.64	0.4410	0.60	0.7390

p-values lower than 0.05 are in bold. *Bearish* (*bullish*) versions of patterns signal a short (long) position

The z-test focuses in each particular case. This means for a given type, version of a pattern and case i, it is being assessed if proportions reported in real and simulated series are equal. Therefore, the null hypothesis is:

$$H_0^i : f_i = \widetilde{f}_i, \tag{5.71}$$

where $i \in \{1,2,3\}$ for cases 1, 2 and 3 respectively. The alternative hypothesis vary from case to case (5.71).

$$H_a^1 : f_1 > \widetilde{f}_1, H_a^2 : f_2 \neq \widetilde{f}_2 \text{ and } H_a^3 : f_3 < \widetilde{f}_3 \tag{5.72}$$

This test is adopted, mainly to examine further the cases where the $H_0^{1,2,3}$ has been rejected. Recall that case 3 represents instances where price does evolved in a desired manner and we close our initial position due to the stop loss conditions imposed whilst case 1 represents instances where price evolves as TA asserts. Therefore, of particular interest are the cases where H_0^1 and H_0^3 were rejected. Such combinations (HS_{Tops}, TT, TB and DB in Table 5.8) imply that after a

Table 5.8 z-Tests for comparing proportions

	z-value	p-value	z-value	p-value
Panel A: HS	HS_{Tops}		$HS_{Bottoms}$	
H_a^1	7.41	**0.0000**	−3.91	1.0000
H_a^2	−0.91	0.3605	14.81	**0.0000**
H_a^3	−4.98	**0.0000**	−6.22	**0.0000**
Panel B: TTTB	*TT*		*TB*	
H_a^1	2.36	**0.0091**	5.83	**0.0000**
H_a^2	2.27	**0.0230**	n/a	n/a
H_a^3	−3.46	**0.0003**	−5.83	**0.0000**
Panel C: DTDB	*DT*		*DB*	
H_a^1	3.51	**0.0002**	7.52	**0.0000**
H_a^2	−1.93	0.0535	−0.34	0.7340
H_a^3	−1.13	0.1286	−7.14	**0.0000**
Panel D: Flags	*Bear*		*Bull*	
H_a^1	1.38	0.0843	−1.65	0.9504
H_a^2	−1.97	**0.0493**	4.82	**0.0000**
H_a^3	0.60	0.7245	−2.27	**0.0115**
Panel E: Pennants	*Bear*		*Bull*	
H_a^1	2.77	**0.0028**	−1.59	0.9445
H_a^2	−1.15	0.2512	1.37	0.1692
H_a^3	−0.30	0.3811	0.19	0.5760
Panel F: Wedges	*Bear*		*Bull*	
H_a^1	−0.48	0.6826	0.32	0.3731
H_a^2	1.26	0.2072	0.60	0.5457
H_a^3	−0.29	0.3857	−0.67	0.2519

p-values in bold highlight statistically significant cases at the 5 % level

technical pattern is confirmed, there are more chances that price evolves as TA asserts in real series than in simulated, $f_1 > \widetilde{f}_1$ and this superiority in the pattern's performance, when real datasets are used, is accompanied by lower proportions of cases 3, $f_3 < \widetilde{f}_3$.

H_0^1 is also rejected in DT and bearish pennants, without accepting the other two alternative hypothesis, H_a^2 and H_a^3. However there are patterns (wedges and bullish pennants) whereby after their formation price evolves similarly in real and simulated series. Finally, flags and $HS_{Bottoms}$, generate significantly different proportions but in cases other than the first one which are not of a technician's primary interest.

5.8 Conclusions

In this chapter, algorithmic methodologies for identifying the bearish and bullish versions of six celebrated technical patterns were proposed and presented. In particular, the proposed mechanisms confirm a pattern's formation when sequences of regional locals fulfill a number of quantitative conditions. These conditions were designed from theoretical descriptions of the examined patterns as found in the literature.

Beginning with the identification of the HS pattern, it was first confirmed that the proposed algorithm successfully identifies the pattern in price series where it had been previously identified via visual assessment. Subsequently, the algorithm was tested on simulated series, generated by the discrete-time version of GBM, with different drift (μ), volatility rates (σ) and rolling windows' widths (w). Surprisingly, the pattern was identified approximately 22 % of the time. This finding, suggests, that since a technical pattern can come into existence out of practically pure randomness, it should also be expected to appear in real series. The relation between the number of patterns identified and various values for μ, σ and w was also examined. It was found that the number of identified HS patterns was positively related with σ and negatively related with w and the absolute value of μ. In addition the choice of w has been discussed and a methodology for defining the width of the RW in the identification process of a zigzag pattern has been proposed.

Three different cases of price evolutions after the realization of a zigzag pattern have been identified; the first case where the price target set by the pattern is reached, the 3rd case where a stop loss condition closes the initial trading position and the second case where none of the above happen until a predefined time limit where the initial position is closed. Relative frequencies of these cases generated from real and simulated series were reported and statistically compared. Briefly, results indicate homogeneity among frequencies that come from real and simulated series for almost half of the examined patterns. However, frequencies regarding case 1 (reaching the price target) were statistically significant greater when obtained from real series rather than simulated ones in 6 out of 12 patterns' versions.

Appendix 1: HS Function

1.1. The Function

$$[PATTERNS] = HS(ys, w, pflag)$$

1.2. Description

This function identifies the HS pattern on a price series by adopting the conditions presented in Sect. 5.2 (Lucke 2003).

Inputs

1. *ys*: Price series (Column Vector).
2. *w*: Is used to define the size of the rolling window which is 2w+1 (see $RW(\cdot)$ in Appendix 1 of Chap. 2)
3. *pflag*: If 1 the function generates a plot.

Outputs

1. *PATTERNS*: is a structured matrix
   - First level: two different sub-structured matrices
     - PATTERNS. NORMAL for HS_{Tops}
     - PATTERNS.INVERSE for $HS_{Bottoms}$
   - Second level: for the normal form (PATTERNS.NORMAL):
     - PATTERNS.NORMAL.Points: This is a $1 \times n$ cell matrix, where n is the number of identified HS_{Tops}. Each cell is a 5×3 matrix containing the characteristic points of the pattern (points P_1, T_1, P_2, T_2 and P_3 of Fig. 5.1). First column corresponds to the x-coordinates, second column takes values of 1 for peaks and 2 for bottoms and third column includes the y-coordinates (prices).
     - PATTERNS.NORMAL.Necklines: A $2 \times n$ matrix containing the slopes (first row) and the constants (second row) for the corresponding necklines.
     - PATTERNS.NORMAL.Breakpoints: A $2 \times n$ matrix containing the x-coordinates (first row) and y-coordinates (second row) of the neckline's penetration.
     - PATTERNS.NORMAL.Numberofnormals: Number of identified HS_{Tops}

       Analogous is the description for the inverse case ($HS_{Bottoms}$).

1.3. Code

```
function [PATTERNS]=HS(ys,w,pflag)
%% Data preparation
l=length(ys);
[Peaks,Bottoms]=RW(ys,w,0);
P_Indx=Peaks(:,2);B_Indx=Bottoms(:,2);
% Peaks take the value of one and bottoms take the value of two.
P_Indx=[P_Indx ones(length(P_Indx),1)];
B_Indx=[B_Indx 2*ones(length(B_Indx),1)];
% Sort locals according to the time observed
PB_Indx=sortrows([P_Indx;B_Indx]);
m=size(PB_Indx,1); %Number of locals
Pot.Normal=[1;2;1;2;1];
Pot.Inverse=[2;1;2;1;2];
Pot.Index=zeros(m,1);
for i=1:m-4
  if PB_Indx(i:i+4,2)==Pot.Normal
    Pot.Index(i+4,1)=1; % Potential Normal form
  elseif PB_Indx(i:i+4,2)==Pot.Inverse
    Pot.Index(i+4,1)=2; % Potential Inverse form
  end
end
PNidx=find(Pot.Index==1);
PIidx=find(Pot.Index==2);
%% HS Tops (Normal Form)
mn=length(PNidx);% Number of potential normal forms
if mn~=0
  Pot.Normalcases.Idx=zeros(mn,1);
  for i=1:mn
    % Examine if conditions [1,3 and 4] are satisfied
    PerCase=zeros(5,3);
    PerCase(:,[1,2])=PB_Indx(PNidx(i,1)-4:PNidx(i,1),:);
    PerCase(:,3)=ys(PerCase(:,1));
    % [i]=Condition i
    if PerCase(3,3)>max(PerCase([1,5],3))&&...%[1]
        PerCase(1,3)>=0.5*sum(PerCase([4,5],3))&& ...%[3]
        PerCase(5,3)>=0.5*sum(PerCase([1,2],3))&& ...%[3]
        PerCase(3,1)-PerCase(1,1)<...
        2.5*(PerCase(5,1)-PerCase(3,1))&&...%[4]
        PerCase(5,1)-PerCase(3,1)<...
        2.5*(PerCase(3,1)-PerCase(1,1))%[4]
      Pot.Normalcases.Idx(i,1)=1;
      Pot.Normalcases.Percases{i,1}=PerCase;
```

```
    end
  end
else Pot.Normalcases.Idx=0;
end
mnn=sum(Pot.Normalcases.Idx);% Number of potential patterns
% after the first filter. It remains only [5] to be checked.
Pot.Normalcases.Idx2=find(Pot.Normalcases.Idx==1);
% Refers to "Pot.Normalcases.Percases"
j=0;%Counter for the PATTERNS matrix
if mnn~=0
  for i=1:mnn%[5]
    NPerCase=...
      Pot.Normalcases.Percases{Pot.Normalcases.Idx2(i,1),1};
    Timelimit=NPerCase(5,1)+NPerCase(5,1)-NPerCase(1,1);
    Neckline.Beta=(NPerCase(4,3)-NPerCase(2,3))/...
      (NPerCase(4,1)-NPerCase(2,1));
    Neckline.Alpha=NPerCase(2,3)-Neckline.Beta*NPerCase(2,1);
    if Timelimit<=l
      Neckline.OU=NPerCase(5,1)+1:Timelimit;
    else
      Neckline.OU=NPerCase(5,1)+1:l;
      % for the cases that the pattern is at the end of the
      % price series
    end
    Neckline.OU=[Neckline.OU;ys(Neckline.OU)'];
    Neckline.OU(3,:)=...
      Neckline.Beta.*Neckline.OU(1,:)+Neckline.Alpha;
    Neckline.OU(4,:)=Neckline.OU(2,:)-Neckline.OU(3,:);
    Neckline.Breakpoint=find(Neckline.OU(4,:)<0,1,'first');
    if isempty(Neckline.Breakpoint)==0
      j=j+1;
      PATTERNS.NORMAL.Points{1,j}=NPerCase;
      PATTERNS.NORMAL.Necklines(1:2,j)=...
        [Neckline.Alpha;Neckline.Beta];
      PATTERNS.NORMAL.Breakpoints(1:2,j)=...
        [Neckline.OU(1,Neckline.Breakpoint);...
        ys(Neckline.OU(1,Neckline.Breakpoint))];
      PATTERNS.NORMAL.Widths(1,j)=...
        NPerCase(4,1)-NPerCase(2,1);
      PATTERNS.NORMAL.Heights(1,j)=(NPerCase(3,3)-...
        (Neckline.Beta*NPerCase(3,1)+Neckline.Alpha));
      PATTERNS.NORMAL.TL(1,j)=PATTERNS.NORMAL.Breakpoints(1,j)+...
        PATTERNS.NORMAL.Widths(1,j);% Time limit
      [~,ystar,~,~]=line_inter(NPerCase([2,4],[1,3]),...
        [PATTERNS.NORMAL.Breakpoints(1,j)-1,...
```

```
        ys(PATTERNS.NORMAL.Breakpoints(1,j)-1);...
        PATTERNS.NORMAL.Breakpoints(1:2,j)']);
      PATTERNS.NORMAL.PT(1,j)=...
        ystar-PATTERNS.NORMAL.Heights(1,j);
    end
  end
end
PATTERNS.NORMAL.Numberofnormals=j;
%% HS Bottoms (Inverse Form)
mi=length(PIidx);% Number of potential inverse forms
if mi~=0
  Pot.Inversecases.Idx=zeros(mi,1);
  for i=1:mi
    %Examine if conditions [1,3 and 4] are satisfied
    PerCase=zeros(5,3);
    PerCase(:,[1,2])=PB_Indx(PIidx(i,1)-4:PIidx(i,1),:);
    PerCase(:,3)=ys(PerCase(:,1));
    if PerCase(3,3)<min(PerCase([1,5],3))&&...%[1]
        PerCase(1,3)<=0.5*sum(PerCase([4,5],3))&& ...%[3]
        PerCase(5,3)<=0.5*sum(PerCase([1,2],3))&& ...%[3]
        PerCase(3,1)-PerCase(1,1)<...
        2.5*(PerCase(5,1)-PerCase(3,1))&&...%[4]
        PerCase(5,1)-PerCase(3,1)<...
        2.5*(PerCase(3,1)-PerCase(1,1))%[4]
      Pot.Inversecases.Idx(i,1)=1;
      Pot.Inversecases.Percases{i,1}=PerCase;
    end
  end
else Pot.Inversecases.Idx=0;
end
mii=sum(Pot.Inversecases.Idx);% Number of potential patterns
% after the first filter. It remains only [5] to be checked.
Pot.Inversecases.Idx2=find(Pot.Inversecases.Idx==1);
% Refers to "Pot.Inversecases.Percases"
jj=0;%Counter for the PATTERNS matrix
if mii~=0
  for i=1:mii%[5]
    NPerCase=...
     Pot.Inversecases.Percases{Pot.Inversecases.Idx2(i,1),1};
    Timelimit=NPerCase(5,1)+NPerCase(5,1)-NPerCase(1,1);
    Neckline.Beta=(NPerCase(4,3)-NPerCase(2,3))/...
      (NPerCase(4,1)-NPerCase(2,1));
    Neckline.Alpha=NPerCase(2,3)-Neckline.Beta*NPerCase(2,1);
    if Timelimit<=l
      Neckline.OU=NPerCase(5,1)+1:Timelimit;
```

```
    else
      Neckline.OU=NPerCase(5,1)+1:l;
      % for the cases that the pattern is at the end of the
      % price series
    end
    Neckline.OU=[Neckline.OU;ys(Neckline.OU)'];
    Neckline.OU(3,:)=...
      Neckline.Beta.*Neckline.OU(1,:)+Neckline.Alpha;
    Neckline.OU(4,:)=Neckline.OU(2,:)-Neckline.OU(3,:);
    Neckline.Breakpoint=find(Neckline.OU(4,:)>0,1,'first');
    if isempty(Neckline.Breakpoint)==0
      jj=jj+1;
      PATTERNS.INVERSE.Points{1,jj}=NPerCase;
      PATTERNS.INVERSE.Necklines(1:2,jj)=...
        [Neckline.Alpha;Neckline.Beta];
      PATTERNS.INVERSE.Breakpoints(1:2,jj)=...
        [Neckline.OU(1,Neckline.Breakpoint);...
        ys(Neckline.OU(1,Neckline.Breakpoint))];
      PATTERNS.INVERSE.Widths(1,jj)=...
        NPerCase(4,1)-NPerCase(2,1);
      PATTERNS.INVERSE.Heights(1,jj)=abs(NPerCase(3,3)-...
        (Neckline.Beta*NPerCase(3,1)+Neckline.Alpha));
  PATTERNS.INVERSE.TL(1,jj)=PATTERNS.INVERSE.Breakpoints(1,jj)+...
        PATTERNS.INVERSE.Widths(1,jj);% Time limit
      [~,ystar,~,~]=line_inter(NPerCase([2,4],[1,3]),...
        [PATTERNS.INVERSE.Breakpoints(1,jj)-1,...
        ys(PATTERNS.INVERSE.Breakpoints(1,jj)-1);...
        PATTERNS.INVERSE.Breakpoints(1:2,jj)']);
      PATTERNS.INVERSE.PT(1,jj)=...
        ystar+PATTERNS.INVERSE.Heights(1,jj);
    end
  end
end
PATTERNS.INVERSE.Numberofinverses=jj;
%% Plotting
if pflag==1 && (j>0 || jj>0)
  plot(ys)
  hold on
  if j>0
    for i=1:PATTERNS.NORMAL.Numberofnormals;
      plot(PATTERNS.NORMAL.Points{1,i}(:,1),...
        PATTERNS.NORMAL.Points{1,i}(:,3),'r*-')
      plot(PATTERNS.NORMAL.Breakpoints(1,i),...
        PATTERNS.NORMAL.Breakpoints(2,i),'rv')
    end
```

```
  end
  if jj>0
    for i=1:PATTERNS.INVERSE.Numberofinverses
      plot(PATTERNS.INVERSE.Points{1,i}(:,1),...
        PATTERNS.INVERSE.Points{1,i}(:,3),'g*-')
      plot(PATTERNS.INVERSE.Breakpoints(1,i),...
        PATTERNS.INVERSE.Breakpoints(2,i),'g^')
    end
  end
  hold off
end
  function [tstar,ystar,LU,LD]=line_inter(A,B)
% A and B are 2x2 matrices with the coordinates of two pairs of
% points that signal two trendlines. Particularly, for flags,
% pennants and wedges points in A and B signal the upper and lower
% trendline respectively. First and second columns include time,
% and price respectively. Trendlines are in the form of:
% y=sx+c, where s is the slope and c is the constant. tstar
% and ystar are coordinates of the point that the two lines intersect.
    S1=(A(2,2)-A(1,2))/(A(2,1)-A(1,1));% slope 1
    S2=(B(2,2)-B(1,2))/(B(2,1)-B(1,1));% slope 1
    C1=A(1,2)-S1*A(1,1);% Constant 1
    C2=B(1,2)-S2*B(1,1);% Constant 2
    tstar=(C2-C1)/(S1-S2);
    ystar=S1*tstar+C1;
    LU=[S1,C1];% Slope and Constant of the upper line
    LD=[S2,C2];% Slope and Constant of the lower line
  end
end
```

Appendix 2: FLAGS Function

2.1. The Function

$$[PATTERNS] = FLAGS(ys, w, typeofdist,\ thor,\ pflag)$$

2.2. Description

This function identifies Flags on a price series by adopting the conditions presented in Sect. 5.4.

Inputs

1. *ys*: Price series (Column Vector).
2. *w*: Is used to define the size of the rolling window which is 2w+1 (see $RW(\cdot)$ in Appendix 1 of Chap. 2)
3. *typeofdist*: See input of $PIPs(\cdot)$ in Appendix 2 of Chap. 2.
4. *thor*: maximum length of the preceding trend (suggested value: 60)
5. *pflag*: If 1 the function generates a plot.

Outputs

1. *PATTERNS*: is a structured matrix
   - First level: two different sub-structured matrices
     - PATTERNS.BULL for bullish flags
     - PATTERNS.BEAR for bearish flags
   - Second level: for the bullish flags (PATTERNS.BULL)
     - PATTERNS.BULL.Points: This is a $1 \times n$ cell matrix, where n is the number of identified patterns. Each cell is a 7×3 matrix containing the characteristic points of the pattern (points $T_{-1}, P_0, T_0, P_1,\ T_1, P_2$ and T_2 of Fig. **5.11**). First column corresponds to the x-coordinates, second column takes values of 1 for peaks and 2 for bottoms and third column includes the y-coordinates (prices).
     - PATTERNS.BULL.TrendLines: This is a $1 \times n$ cell matrix, where each cell is a 2×2 matrix containing the slopes (first column) and the constants (second column) for the corresponding trendlines.
     - PATTERNS.BULL.Breakpoints: A $2 \times n$ matrix containing the x-coordinates (first row) and y-coordinates (second row) of the upper trend line's penetration.
     - PATTERNS.BULL.Numberofpatterns: Number of identified bullish flags

       Analogous is the description for the bearish flags.

2.3. Code

```
function [PATTERNS]=FLAGS(ys, w, typeofdist, thor, pflag)
%% Data preparation
l=length(ys);
[Peaks,Bottoms]=RW(ys,w,0);
% Ignore locals identified the first days
Peaks(Peaks(:,2)<=20,:)=[];
Bottoms(Bottoms(:,2)<=20,:)=[];
P_Indx=Peaks(:,2);B_Indx=Bottoms(:,2);
P_Indx=[P_Indx ones(length(P_Indx),1)];
```

```
B_Indx=[B_Indx 2*ones(length(B_Indx),1)];
PB_Indx=sortrows([P_Indx;B_Indx]);
m=size(PB_Indx,1); % Number of locals
Pot.FlagBull=[1;2;1;2];
Pot.FlagBear=[2;1;2;1];
Pot.Index=zeros(m,1);
for i=1:m-3
  if PB_Indx(i:i+3,2)==Pot.FlagBull
    Pot.Index(i+3,1)=1; % Potential Bullish Flags
  elseif PB_Indx(i:i+3,2)==Pot.FlagBear
    Pot.Index(i+3,1)=2; % Potential Bearish Flags
  end
end
%% FLAGS (BULLISH)
PFBullidx=find(Pot.Index==1);
j=0;% Counter
mn=length(PFBullidx);
if mn~=0
  for i=1:mn
    PC=zeros(4,3);% PerCase
    PC(:,[1,2])=PB_Indx(PFBullidx(i,1)-3:PFBullidx(i,1),:);
    PC(:,3)=ys(PC(:,1));
    YS=[ys,(1:l)'];
    if PC(1,1)-thor<=0
      Flocalx=1;
    else
      Flocalx=PC(1,1)-thor;
    end
    YS2=YS(Flocalx:PC(1,1),:);
    % Beginning of the prevailing trend
    [~,Flocalx2]=min(YS2(:,1));
    YS3=YS2(Flocalx2:end,:);
    if size(YS3,1)<4
      YS3=YS2(Flocalx2-4:end,:);
    end
    [PIPxy]=PIPs(YS3(:,1),4,typeofdist,0);
   % PIPs identification
    PCadd=[YS3(PIPxy(1:3,1),2),[2;1;2],PIPxy(1:3,2)];
    PC=[PCadd;PC];% Add 3 rows to PC (with PIPS)

[tstar,~,LU,LD]=line_inter(PC([4,6],[1,3]),PC([5,7],[1,3]));
    tl=2*PC(7,1)+PC(4,1);%Time limit for penetration
    if abs(tl)>l,tl=l;end
   % In case time limit exceeds the ys length
    Tlines=[LU(1).*(PC(7,1)+1:tl)+LU(2);...
```

```
      LD(1).*(PC(7,1)+1:tl)+LD(2)];
    Y=[ys(PC(7,1)+1:tl)';ys(PC(7,1)+1:tl)'];
    OU=Y-Tlines>=0;
    Utb=PC(7,1)+find(OU(1,:)==1,1,'first');
   % Upward penetration
    Dtb=PC(7,1)+find(OU(2,:)==0,1,'first');
   % Downward penetration
    penconf=0;% Upward penetration confirmation
    if isempty(Utb)==0%[5,6]
      if isempty(Dtb)==1 && Utb-PC(4,1)<=20
        penconf=1;
      elseif isempty(Dtb)==0 && Utb<Dtb && Utb-PC(4,1)<=20
        penconf=1;
      end
    end
    [uly ulx]=max(PC([2,3],3));ulx=PC(1+ulx,1);%[2]
    if PC(4,3)>uly &&...%[1]
        uly< PC(5,3)+((ulx-PC(5,1))*...
        (PC(7,3)-PC(5,3))/(PC(7,1)-PC(5,1))) &&...%[2]
        PC(6,3)<PC(4,3) && PC(7,3)<PC(5,3) &&...%[3]
        abs(tstar-(PC(4,1)+PC(7,1))/2)>...
        5*(PC(7,1)-PC(4,1)) && penconf==1 &&...%[4,5]
        (PC(4,1)-PC(1,1))>2*(Utb-PC(4,1))%[7]
      j=j+1;
      PATTERNS.BULL.Points{1,j}=PC;
      PATTERNS.BULL.TrendLines{1,j}=[LU;LD];
      PATTERNS.BULL.Breakpoints(1:2,j)=[Utb;ys(Utb)];
      PATTERNS.BULL.Widths(1,j)=Utb-PC(4,1);
      [~,ystar,~,~]=line_inter(PC([4,6],[1,3]),...
        [Utb-1,ys(Utb-1);Utb,ys(Utb)]);
      PATTERNS.BULL.PT(1,j)=ystar+PC(4,3)-PC(1,3);
      PATTERNS.BULL.TL(1,j)=Utb+PC(4,1)-PC(1,1);
    end
  end
end
PATTERNS.BULL.Numberofpatterns=j;
%% FLAGS (BEARISH)
PFBearidx=find(Pot.Index==2);
j=0;% Counter
mn=length(PFBearidx);
if mn~=0
  for i=1:mn
    PC=zeros(4,3);% PerCase
    PC(:,[1,2])=PB_Indx(PFBearidx(i,1)-3:PFBearidx(i,1),:);
    PC(:,3)=ys(PC(:,1));
```

```
    YS=[ys,(1:l)'];
    if PC(1,1)-thor<=0
      Flocalx=1;
    else
      Flocalx=PC(1,1)-thor;
    end
    YS2=YS(Flocalx:PC(1,1),:);
    [~,Flocalx2]=max(YS2(:,1));
    YS3=YS2(Flocalx2:end,:);
    if size(YS3,1)<4
      YS3=YS2(Flocalx2-4:end,:);
    end
    [PIPxy]=PIPs(YS3(:,1),4,typeofdist,0);
    PCadd=[YS3(PIPxy(1:3,1),2),[2;1;2],PIPxy(1:3,2)];
    PC=[PCadd;PC];

[tstar,~,LU,LD]=line_inter(PC([5,7],[1,3]),PC([4,6],[1,3]));
    tl=2*PC(7,1)+PC(4,1);%Time limit for penetration
    if abs(tl)>l,tl=l;end
    Tlines=[LU(1).*(PC(7,1)+1:tl)+LU(2);...
      LD(1).*(PC(7,1)+1:tl)+LD(2)];
    Y=[ys(PC(7,1)+1:tl)';ys(PC(7,1)+1:tl)'];
    OU=Y-Tlines>=0;
    Utb=PC(7,1)+find(OU(1,:)==1,1,'first');
   % Upward penetration
    Dtb=PC(7,1)+find(OU(2,:)==0,1,'first');
   % Downward penetration
    penconf=0;% Downward penetration confirmation
    if isempty(Dtb)==0%[5,6]
      if isempty(Utb)==1 && Dtb-PC(4,1)<=20
        penconf=1;
      elseif isempty(Utb)==0 && Dtb<Utb && Dtb-PC(4,1)<=20
        penconf=1;
      end
    end
    [lly llx]=min(PC([2,3],3));llx=PC(1+llx,1);%[2]
    if PC(4,3)<lly &&...%[1]
        lly< PC(5,3)+((llx-PC(5,1))*...
        (PC(7,3)-PC(5,3))/(PC(7,1)-PC(5,1))) &&...%[2]
        PC(6,3)>PC(4,3) && PC(7,3)>PC(5,3) &&...%[3]
        abs(tstar-(PC(4,1)+PC(7,1))/2)>...
        5*(PC(7,1)-PC(4,1)) && penconf==1 &&...%[4,5]
        (PC(4,1)-PC(1,1))>2*(Dtb-PC(4,1))%[7]
      j=j+1;
      PATTERNS.BEAR.Points{1,j}=PC;
```

```
      PATTERNS.BEAR.TrendLines{1,j}=[LU;LD];
      PATTERNS.BEAR.Breakpoints(1:2,j)=[Dtb;ys(Dtb)];
      PATTERNS.BEAR.Widths(1,j)=Dtb-PC(4,1);
      [~,ystar,~,~]=line_inter(PC([4,6],[1,3]),...
        [Dtb-1,ys(Dtb-1);Dtb,ys(Dtb)]);
      PATTERNS.BEAR.PT(1,j)=ystar+PC(4,3)-PC(1,3);
      PATTERNS.BEAR.TL(1,j)=Dtb+PC(4,1)-PC(1,1);
    end
  end
end
PATTERNS.BEAR.Numberofpatterns=j;
%% Plotting
if pflag==1 && (j>0 || PATTERNS.BULL.Numberofpatterns>0)
  figure(1)
  plot(ys),hold on
  for i=1:PATTERNS.BULL.Numberofpatterns
    plot(PATTERNS.BULL.Points{1,i}(:,1),...
      PATTERNS.BULL.Points{1,i}(:,3),'g')
    plot(PATTERNS.BULL.Points{1,i}([4,6],1),...
      PATTERNS.BULL.Points{1,i}([4,6],3),'k--')
    plot(PATTERNS.BULL.Points{1,i}([5,7],1),...
      PATTERNS.BULL.Points{1,i}([5,7],3),'k--')
    plot(PATTERNS.BULL.Breakpoints(1,:),...
      PATTERNS.BULL.Breakpoints(2,:),'g^')
  end
  for i=1:PATTERNS.BEAR.Numberofpatterns
    plot(PATTERNS.BEAR.Points{1,i}(:,1),...
      PATTERNS.BEAR.Points{1,i}(:,3),'r')
    plot(PATTERNS.BEAR.Points{1,i}([4,6],1),...
      PATTERNS.BEAR.Points{1,i}([4,6],3),'k--')
    plot(PATTERNS.BEAR.Points{1,i}([5,7],1),...
      PATTERNS.BEAR.Points{1,i}([5,7],3),'k--')
    plot(PATTERNS.BEAR.Breakpoints(1,:),...
      PATTERNS.BEAR.Breakpoints(2,:),'rv')
  end
  hold off
end
end
```

References

Achelis SB (1995) Technical analysis from A to Z. Probus Publishing, Chicago
Bry G, Boschan C (1971) Cyclical analysis of time series: selected procedures and computer programs. Technical paper 20. National Bureau of Economic Research, New York
Bulkowski TN (2000) Encyclopedia of chart patterns. Wiley, New York
Bulkowski T (2002) Trading classic chart patterns. Wiley, New York
Chang PHK, Osler CL (1999) Methodical madness: technical analysis and the irrationality of exchange-rate forecasts. Econ J 109:636–661
Clyde WC, Osler CL (1997) Charting: Chaos theory in disguise? J Futur Mark 17(5):489–514
Dawson ER, Steeley JM (2003) On the existence of visual technical patterns in the UK stock market. J Bus Financ Account 30(1 and 2):263–293
De Bondt W (1998) A portrait of the individual investor. Eur Econ Rev 42(3–5):831–844
Edwards RD, Magee J (1997) Technical Analysis of Stock Trends, 7th edn. John Magee, Inc., Boston
Gilovich T (1993) How we know what Isn't so. Free Press, New York
Hull JC (2006) Options, futures, and other derivatives, 6th edn. Prentice Hall, New Jersey
Levy RA (1971) The predictive significance of five-point chart patterns. J Bus 44(3):316–323
Lo AW, Mamaysky H, Wang J (2000) Foundations of technical analysis: computational algorithms, statistical inference, and empirical implementation. J Financ 55(4):1705–1765
Lucke B (2003) Are technical trading rules profitable? Evidence for head-and-shoulder rules. Appl Econ 35:33–40
Murphy J (1986) Technical analysis of the future markets: a comprehensive guide to trading methods and applications. Prentice Hall, New York
Murphy J (1999) Technical analysis of the financial markets: a comprehensive guide to trading methods and applications. New York Institute of Finance, New York
Neftci SN (1991) naïve trading rules in financial markets and Wiener-Kolmogorov prediction theory: a study of "Technical Analysis". J Bus 64(4):549–571
Osler CL, Chang PHK (1995) Head and shoulders: not just a flaky pattern. Federal Reserve Bank of New York Staff Reports, Report No. 4.
Pring MJ (2002) Technical analysis explained: the successful investor's guide to spotting investment trends and turning points, 4th edn. McGraw-Hill Book Company, New York
Roberts H (1959) Stock-market "Patterns" and Financial analysis: methodological suggestions. J Financ 14(1):1–10
Sklarew A (1980) Techniques of a professional commodity chart analyst. Commodity Research Bureau, New York
Warneryd K-E (2001) Stock-market psychology. Edward Elgar, Cheltenham
Zapranis A, Tsinaslanidis P (2010a) A behavioral view of the head-and-shoulders technical analysis pattern. Paper presented at the Proceedings of the 3rd International Conference in Accounting and Finance, Skiathos, Greece
Zapranis A, Tsinaslanidis P (2010b) Identification of the head-and-shoulders technical analysis pattern with neural networks. Paper presented at the artificial neural networks - ICANN 2010, 20th International Conference, Thessaloniki, Greece, September 15–18, 2010
Zielonka P (2004) Technical analysis as the representation of typical cognitive biases. Int Rev Financ Anal 13(2):217–225

Chapter 6
Circular Patterns

6.1 Introduction

Rounding Bottoms/Tops (RB/RT) is another celebrated technical pattern. Due to its circular form, RB is also known as saucer bottom or "U" formation. The pattern is formed when market's expectations gradually shift from bearish to bullish (Achelis 1995). The essence of this pattern is that prices gradually decline up to a specific point where a regional bottom is created, and after this point the trend changes to "bullish". This evolution of the price creates a "bowl-shaped" pattern, with the two "leaks" of the "bowl" signifying a resistance level. After the upwards penetration of this level prices are expected to rise in a level equal with the patterns depth. The volume during the creation of the pattern is similar. RB is an indicative example of circular technical patterns. RT (or inverse saucers) is the mirrored image of RB and it develops at a market peak. Both patterns result by gradual changeover in the demand and supply balance that slowly picks up momentum in the direction opposite to that of the previous trend (Pring 2002).

This chapter focuses on the proposition of a novel algorithmic approach for identifying these two versions of the pattern. Similarly with the methodologies presented in Chaps. 4 and 5, the uniqueness of the proposed methodology resides in the manner that conventions of TA are incorporated in the identification mechanism. To be more precise, from different theoretical descriptions of the RB, and with simple geometric rules, a new rule-based identification mechanism is developed which extracts a great proportion of subjectivity included in technical pattern recognition process via visual assessment.

The advantage of this method is that it is simple and it can identify simultaneously patterns of different size (both width and depth) where in case of previous studies (Wang and Chan 2009) which use template-matching techniques the user has to define the size of the template window before the identification process starts. The template-matching is described in Duda and Hart (1973) and is used in various studies that evaluate the profitability of various technical patterns like variations of

P.E. Tsinaslanidis, A.D. Zapranis, *Technical Analysis for Algorithmic Pattern Recognition*, DOI 10.1007/978-3-319-23636-0_6

bull flags and volume spikes (Leigh et al. 2002a, b, c, 2004, 2005, 2008; Bo et al. 2005; Wang and Chan 2007, 2009). The identification methodology presented in this chapter is the one initially proposed in Zapranis and Tsinaslanidis (2012).

The rest of the chapter is organized as follows. In Sect. 6.2 the identification mechanism for RB and RT patterns is presented. Three possible price evolutions following a pattern's fully formation are defined in a manner similar to the one used in Chap. 5. These distinct price behaviors are statistically assessed in Sect. 6.3, where a comparison between results obtained from real and artificial created series is made. Finally, in Sect. 6.4 we conclude. In Appendix 1 the developed Matlab code for RB pattern's identification is provided.

6.2 Identifying Rounding Tops/Bottoms

Bibliography focusing on the recognition of saucers is limited. The explanation for this is that its identification is generally considered as a difficult task to implement. More precisely, Murphy (1986) states that "*There are no precise measuring rules for the saucer bottom*". A similar statement is found in Edwards and Magee (1997) where they mention that "*It is difficult to set precise rules for trading on these gradual changes of trend*". Regarding the theoretical identification "guidelines" found in bibliography, there is evidence that the evolution of the price is gradual and should form a semicircle. Again, referring to Murphy (1986) the following excerpt is found: "*This pattern represents a very slow and gradual change in trend*". Achelis (1995) confirms by stating that "*Rounding bottoms occur as expectations gradually shift from bearish to bullish*". Finally, according to Pring (2002), "*A saucer is constructed by drawing a circular line under the lows...*". Considering the aforementioned pattern's descriptions emphasizing the circular characteristics of RB and RT patterns the following identification mechanism is proposed.

Initially, an optimal case of the saucers pattern is considered and then more flexibility is added. For the pattern's identification, three major characteristic points are necessary: two points at about the same level (P_1 and P_2) and a local trough (T_1) approximately in the middle of them. In an ideal case (Fig. 6.1) points P_1 and P_2 would be at the exact same level, and the local bottom would be at the midpoint of them. The general rule implemented in this process is that the vast majority of closing prices between the points (P_1 and P_2) must fluctuate, within specific circular boundaries; around the lower half part of the circumscribed circle (semicircle) defined by the isosceles triangle $P_1T_1P_2$.

In an ideal saucers case the three critical points form an isosceles[1] triangle (Fig. 6.2). For simplicity the triangle $P_1T_1P_2$ is denoted as $\Delta(ABC)$. The sides of the triangle are noted with small caps (a, b and c). From trigonometry it is known

[1] An isosceles triangle is any triangle that has exact two sides of the same length.

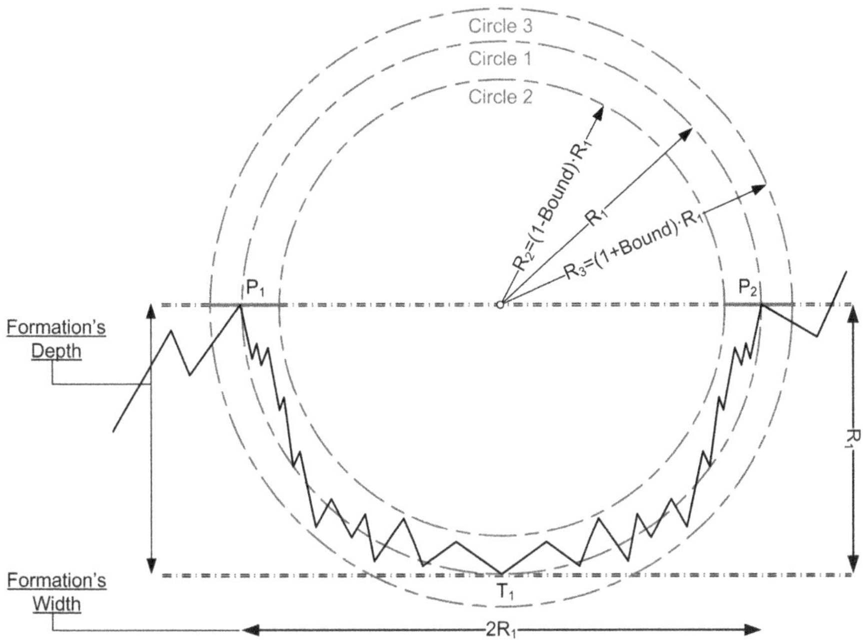

Fig. 6.1 An ideal case of RB pattern. A price series must fluctuate within the bounds set by the two homocentric circles (*circle 2* and *3*). In this ideal case the depth of the pattern equals with the radius (R_1) of the *circle 1* and the width with two radius ($2R_1$) giving a scale of 1:2. P_1, T_1 and P_2 are the pattern's characteristic points

that every triangle has a circumscribed circle,[2] which center (K) is the meeting point of the triangle's three perpendicular bisectors.[3] The radius (R_1) of the circumscribed circle is given by (6.1).

$$R_1 = \frac{a}{2\sin(A)} \tag{6.1}$$

Here, a is the triangle's side opposite of the vertex A and sin is the sine (Fig. 6.2).

At this point the aforementioned triangle is placed on a Cartesian coordinate system in two dimensions. Thus, each point would be defined by a pair of coordinates; the time and the stock's closing price. Given the coordinates of two points, the distance (D) between these two points is given by (6.2).

[2] A circumscribed to a triangle circle is a circle passing through all three triangle's vertices.

[3] A perpendicular bisector of a triangle's side is a straight line passing through the side's midpoint and being perpendicular to it.

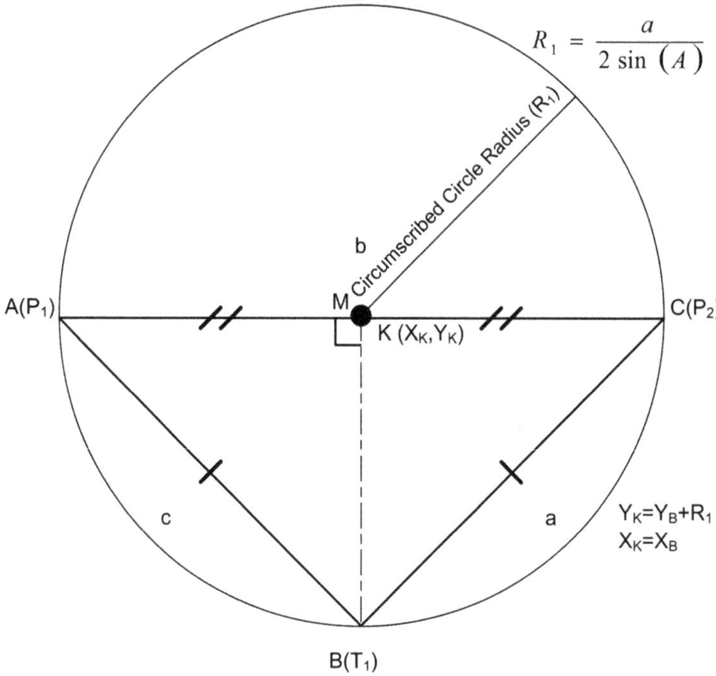

Fig. 6.2 Estimating the radius (R_1) and the center's coordinates $K(X_K, Y_K)$

$$D = \sqrt{(dx)^2 + (dy)^2} \tag{6.2}$$

In (6.2), $dx(dy)$ is the difference between the x-coordinates (y-coordinates) of these two points. Conditioning that the coordinates (time and stock price) for points B and C are known, the calculation of the length (or distance) for the side a is feasible. Also sin(A) equals with the side BM divided by the side AB (or c). But $c = a$ and side BM equals with the difference of the y-coordinates (prices) of points A (or C) with B. From the above analysis the radius (R_1) of the circumscribed circle is calculated and the coordinates of the circumscribed circle's center (point K in Fig. 6.2) result. To be more precise, if (X_K, Y_K) and (X_B, Y_B) are the coordinates of the points K and B respectively, then $X_K = X_B$ and $Y_K = Y_B + R_1$.

So far the radius (R_1) and the center of the circumscribed circle have been determined. The next step is the calculation of the corresponding y-coordinates ($\hat{y}_{i,j}$) for all the available trading days (x_i) between points A and C, for the three $j \in \{1, 2, 3\}$ homocentric circles presented in Fig. 6.1. To do so, the second order equation presented in (6.3) is solved. Points with coordinates $(x_i, \hat{y}_{i,1})$, correspond to the circle's points that the stock price should fluctuate on in the case of a RB pattern (circle 1 in Fig. 6.1). Pattern's confirmation occurs when prices are generally constrained within the bounds set by two other homocentric circles (circles 2 and 3 in Fig. 6.1). The bounds used are expressed as a percentage of the radius of the circle 1. More precisely, if a bound rate of 20 % is used then circles 2 and 3 would have a radius of $R_2 = 0.8R_1$ and $R_3 = 1.2R_1$ respectively. $(x_i, \hat{y}_{i,2})$ and $(x_i, \hat{y}_{i,3})$ signify the bounds in where the vast majority of prices should fluctuate in order to realize a successful pattern's identification.[4]

$$R_j = \sqrt{(dx)^2 + (dy)^2} = \sqrt{(x_i - K_x)^2 + \left(\hat{y}_{i,j} - K_y\right)^2} \\ \Leftrightarrow \hat{y}_{i,j}^2 - 2K_y\hat{y}_{i,j} + K_y^2 - R_j^2 + (x_i - K_x)^2 = 0, \forall i \in \mathbb{Z} \cap [A_x, C_x] \tag{6.3}$$

At this point, the manner by which the algorithm seeks for the three pattern's characteristic points (P_1, T_1 and P_2) has to be specified. Before the initial downtrend a short term local peak (P_1) must occur. By implementing the $RW(\cdot)$ (Chap. 2), short term local peaks are being identified. Each identified peak sets a horizontal price level and subsequently we trace the first forthcoming day that stock's price overcomes this level (point P_2). In an ideal pattern these two points would have the same value but in reality this situation is extreme rare. What it is for sure is that these two points would be at about the same level and should also have a local minimum between them (T_1). There is no need to check if this point is at the midpoint between P_1 and P_2, because it is also extreme rare to be at exact the half distance between the two "leaks" of the "bowl".

The next step is a scaling process that has to take place, in order to simulate the way a technical analyst seeks for a pattern in a price series. It can be argued, that in such case the investor actually focuses on specific parts of the price series he is examining, ignoring the actual price level of the stock. In addition, he sets the price series in a system of coordinates and scales instinctively the data into his mind. It is of prime importance the scaling applied in every potential saucers pattern. In the ideal case presented in Fig. 6.1, it is asserted that the pattern's width equals with two radiuses ($2R_1$) and the depth of the pattern equals with one radius (R_1). Thus, for the calculation of the semicircles, and for a proper identification process, a half size, "depth/width" scale (1/2) is necessary, for every potential pattern. After identifying these three characteristic points and scaling the data, it is very important to set the coordinates of points A, B and C to construct the isosceles triangle $\Delta(ABC)$ in Fig. 6.2. In an ideal case, scaled points $\widetilde{P}_1, \widetilde{T}_1$ and $\widetilde{P}_2$ of a stock price series would be the same with points A, B, and C of Fig. 6.2 and would form an isosceles right triangle. However, as already argued, in reality this is very difficult to happen. So

[4] Solving (6.3) gives two solutions for every x_i. In case of RB (RT), the minimum (maximum) value of each pair of solutions is kept for drawing the homocentric semicircles.

Table 6.1 Adjusting the coordinates of the candidate patterns' characteristic points

	Points used for the creation of the semicircles	
Scaled points observed in the stock	x-Coordinates	y-Coordinates
$\widetilde{P}_1\left(x_{\widetilde{P}_1}, y_{\widetilde{P}_1}\right)$	$x_A = x_{\widetilde{P}_1}$	$y_A = \left(y_{\widetilde{P}_1} + y_{\widetilde{P}_2}\right)/2$
$\widetilde{T}_1\left(x_{\widetilde{T}_1}, y_{\widetilde{T}_1}\right)$	$x_B = \left(x_{\widetilde{P}_1} + x_{\widetilde{P}_2}\right)/2$	$y_B = y_{\widetilde{T}_1}$
$\widetilde{P}_2\left(x_{\widetilde{P}_2}, y_{\widetilde{P}_2}\right)$	$x_C = x_{\widetilde{P}_2}$	$y_C = \left(y_{\widetilde{P}_1} + y_{\widetilde{P}_2}\right)/2$

Formulas used to covert scaled coordinates of points $\widetilde{P}_1$, $\widetilde{T}_1$ and $\widetilde{P}_2$ into points A, B and C. The coordinates of the latest points form an isosceles triangle on a Cartesian system in two dimensions. The semicircles drawn for the identification of the saucers pattern are based on this triangle.

points A and C will keep the x-coordinates of points $\widetilde{P}_1$ and $\widetilde{P}_2$, but as far as their y-coordinates is concerned, they will have the average of their y-coordinates. Finally, point B will have the average of the x-coordinates of the points $\widetilde{P}_1$ and $\widetilde{P}_2$, and the y-coordinates of point $\widetilde{T}_1$. Table 6.1 presents these adjustments. Subsequently, from the $\Delta(ABC)$ the three homocentric circles are going to be drawn and it is going to be assessed whether all daily adjusted closing prices, including those of points $\widetilde{P}_1$, $\widetilde{T}_1$ and $\widetilde{P}_2$, fluctuate within the borders set by these semicircles. In such confirmation a saucers pattern would be successfully identified.

That way, the proposed rule-based mechanism is able to identify patterns of different size at once. This is not the case in template-matching techniques were different sizes of template's grid are used every time. In order to avoid cases with just a few observations, a width-threshold (t_{width}) of 15 days is employed. This infers that the pattern must have at least 15-days width (three trading weeks). The threshold of 15 days is not set arbitrarily but by taking into consideration what Pring (2002) reports:

> Rounding and saucer formations... can take as little as 3 weeks to as long as several years to complete.

An additional threshold is adopted (t_{Fit}) in order to specify the minimum proportion of observations fluctuating between the estimated bounds. The fit value of a RB (or RT) pattern is calculated by (6.4).

$$Fit = \frac{Number\ of\ observations\ fluctuating\ within\ bounds}{Total\ number\ of\ pattern's\ observations} \tag{6.4}$$

Finally, Pring (2002) mentions:

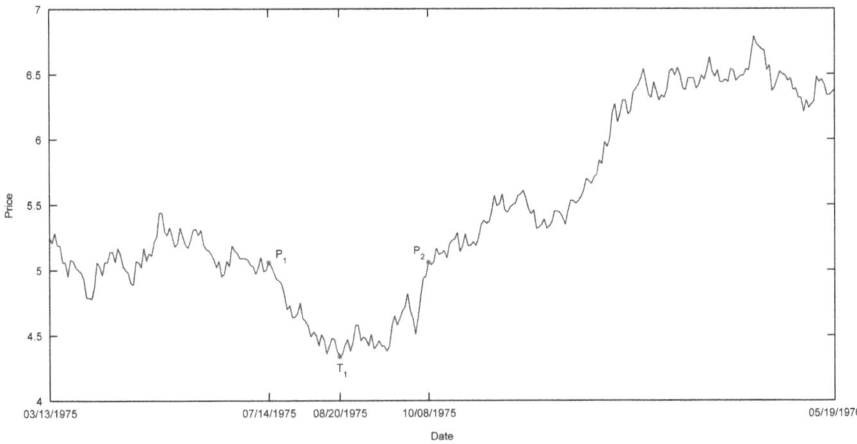

Fig. 6.3 Successful identification of the RB pattern on the adjusted closing prices of IBM stock

The significance of a price formation or pattern is a direct function of its size and depth.

The depth of each pattern (6.5) is calculated in a manner similar to that presented in Wang and Chan (2009).

$$Depth = \frac{P_{max} - P_{min}}{P_{min}} \tag{6.5}$$

In RB case, $P_{max} = \max(y_{P_1}, y_{P_2})$ and $P_{min} = y_{T_1}$, whilst in RT case $P_{max} = y_{P_1}$ and $P_{min} = \min(y_{T_1}, y_{T_2})$.

The aforementioned RB identification is performed by the developed *RBottoms* $(\cdot)$ (see Appendix 1).[5]

Figure 6.3 presents a case of a 62-days saucers pattern identified in the IBM stock at the period 14-Jul-1975 to 8-Oct-1975. In Fig. 6.4, the corresponding part of the price series is being isolated and leveled with a 1/2 scale. After that, the homocentric semicircles with a bound level of 20 % are being drawn. 59 out of 62 closing prices fluctuate within the pre-specified bounds, constituting a fit value of 95.16 %.

Similarly, a RT example is illustrated in Figs. 6.5 and 6.6 for the Oracle stock. A 23-days RT is identified in the period 1-Apr-2010 to 4-May-2010. The fit value is this case is 91.3 % resulted by 21 out of 23 days fluctuating within the predefined bounds.

[5] For brevity reasons only the *RBottoms*$(\cdot)$ is provided, since the identification of RT is a mirror process of the identification of RB.

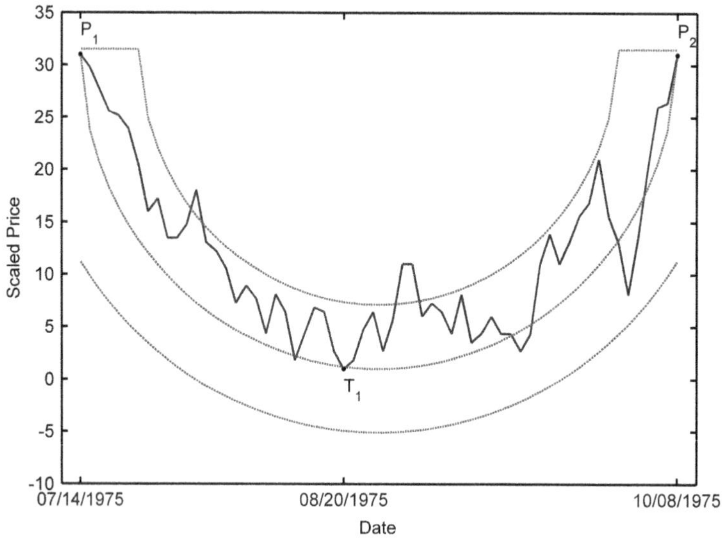

Fig. 6.4 RB identification in scaled data (scale of 1/2). Focusing on the saucers pattern identified in IBM stock price series (Fig. 6.3). Pattern's parameters: Width = 62, Depth = 16.59 %, Fit = 95.16 %

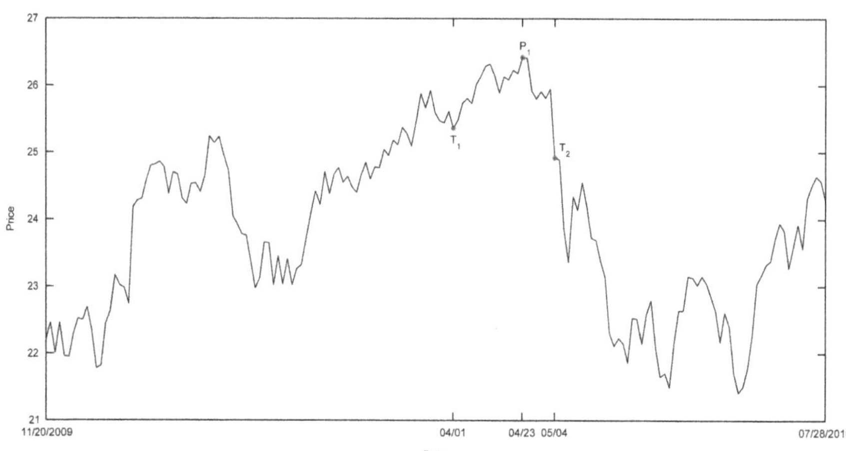

Fig. 6.5 Successful identification of the RT pattern on the adjusted closing prices of Oracle stock

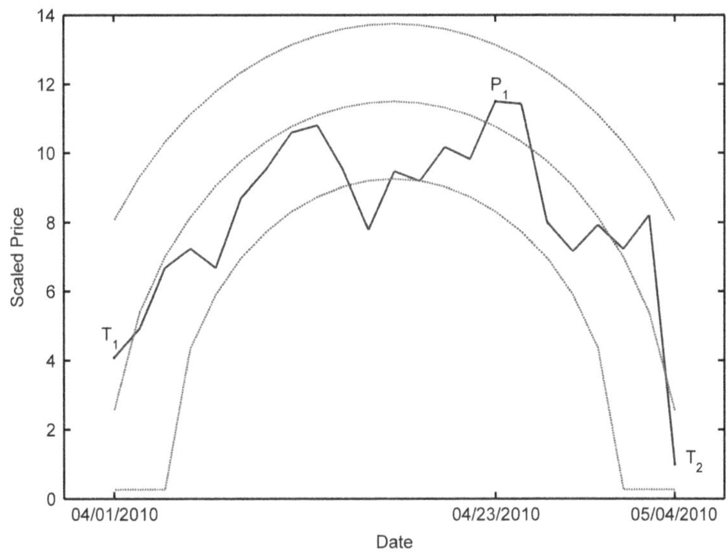

Fig. 6.6 RT identification in scaled data (scale of 1/2). Focusing on the saucers pattern identified in Oracle stock price series (Fig. 6.5). Pattern's parameters: Width = 23, Depth = 6.02 %, Fit = 91.3 %

6.3 Assessing the Predictive Performance

This section assesses the predictive performance of RB and RT patterns adopting a similar methodology with the one presented in Chap. 5. In particular three mutually exclusive cases are also defined that describe the manner that a price series evolves after a pattern's confirmation. In the first case, instances where prices reach a predefined *price target* (PT) within a *time limit* (TL) are classified. For the RB (RT), the price target is set by adding (subtracting) the depth of the pattern to the maximum (minimum) "leak" of the pattern, i.e., for the RB it is $PT = P_{max} + depth \times P_{mim}$ whilst for the RT, $PT = P_{min} - depth \times P_{mim}$. The TL is defined by the pattern's width in the same manner as it was defined in the zig-zag patterns. A stop loss condition is also applied by adopting two thresholds, t_n and t_{sl}. More precisely, t_{sl} is the maximum loss accepted after opening a position and t_n is the maximum number of consecutive days where the price breaches the resistance (support) level defined by the first local of the RB (RT) pattern. When TL expires without prices reaching the PT or fulfilling the stop loss conditions, the instance is classified as second case. Finally, the third case includes occurrences where the stop loss conditions are fulfilled.

The algorithm presented in Sect. 6.2 was adopted on the same dataset used in Chaps. 4 and 5 (232 and 501 stocks composing the NYSE and NASDAQ index respectively). Tables 6.2 and 6.3 present the results from the NYSE and NASDAQ components respectively. It has found that the RB/RT patterns were identified at least one time in all price series considered which was expected due to the large

Table 6.2 RB and RT results from NYSE components

Case (i)	$f_i(\%)$	$r_i(\%)$	$r_{i,daily}(\%)$
Panel A: RB	$n_p = 8{,}792, n_s = 501$		
1	15.45	12.16	0.61
2	6.21	4.97	0.17
3	78.34	−2.46	−0.35
Total	100.00	0.26	0.02
Panel B: RT	$n_p = 6{,}833, n_s = 501$		
1	7.05	16.21	1.01
2	6.59	9.15	0.25
3	86.36	−3.01	−0.41
Total	100.00	−0.85	−0.09

$Bounds = 0.3$, $tWidth = 15$, $t_{Fit} = 0.9$, $w = 10$, $t_n = 2$ and $t_{sl} = -0.04$

i case 1, 2 or 3 describing the price evolution after identification of each pattern, $f_i(\%)$ relative frequency of each case, $r_i(\%)$ average return for each case, $r_{i,daily}(\%)$ average daily return for each case, n_p number of identified patterns in original series, n_s number of original price series where at least one pattern was identified

Table 6.3 RB and RT results from NASDAQ components

Case (i)	$f_i(\%)$	$r_i(\%)$	$r_{i,daily}(\%)$
Panel A: RB	$n_p = 4{,}632, n_s = 232$		
1	14.21	18.97	0.97
2	3.93	7.20	0.22
3	81.87	−4.10	−0.62
Total	100.00	−0.38	−0.04
Panel B: RT	$n_p = 2{,}660, n_s = 232$		
1	5.83	19.03	1.45
2	7.33	16.54	0.40
3	86.84	−4.13	−0.67
Total	100.00	−1.26	−0.14

$Bounds = 0.3$, $tWidth = 15$ $t_{Fit} = 0.9$, $w = 10$, $t_n = 2$ and $t_{sl} = -0.04$

As for Table 6.2

length of the time period considered. A common finding in both datasets and in both pattern's versions is that proportions of cases where price meets the predefined PT (case 1) are significantly small. For example, most of the times $f_3 \gg f_1 > f_2$ which is consistent with the findings reported in Chap. 5 for the zigzag patterns. This implies that also these circular patterns do not perform as TA asserts most of the time. Positive returns were generated from cases 1 which are greater than the positive returns generated of cases 2 and the negative returns of cases 3. Overall, $r_1 > r_2 > 0 > r_3$ holds which is also aligned with the findings reported for the zigzag patterns. Finally, the great positive returns generated by successful cases are not sufficient to compensate the negative returns generated when a stop loss condition is adopted. Thus the average realised return is negative, with an exception of the one generated by RB in NYSE stocks.[6]

Subsequently, proportions, f_i, of each case i, $i \in \{1, 2, 3\}$, resulted from the actual series are compared with those generated from simulated series, $\widetilde{f}_i$. Simulated series were generated by using GBM (5.14). In particular, it is examined if the reported price behaviour after a pattern's confirmation is similar in real and simulated price series. This examination has been carried by adopting two statistical tests; x^2-test for homogeneity of proportions and z-test for comparing proportions. The simulation process and the null and alternative hypothesis tested by the aforementioned tests are the same with those presented in Chap. 5. In Table 5.6 it is shown that x^2-tests rejected the null hypothesis ($H_0^{1,2,3}$) in both versions of the examined pattern and in both datasets at a significance level of 5 %. This implies that the allocation of price behaviour to the three different cases in real and simulated series in not similar. Subsequently, z-tests were carried to identify the source of this heterogeneity, with a significance level of 5 %. More precisely regarding the RT pattern, in both datasets H_0^1 was rejected which implies that after the RT formation, price reaches the PT more frequently when actual series are used i.e., $f_1 > \widetilde{f}_1$. However, H_a^3 was not accepted at the same time, since stop loss conditions are also met more frequently in the actual series. In fact, $f_1 > \widetilde{f}_1$ is attributed to the lower proportions reported in cases 2, $f_2 < \widetilde{f}_2$. On the other hand, RB pattern did perform better in NYSE stocks, as all z-tests rejected the null hypothesis, which implies that in real series price reached the PT more frequently (and at the same time stop loss conditions were met less frequently) than it did in simulated series. Finally, in NASDAQ series, the RB pattern did present statistically significant lower proportions of cases 3 in real series, however the proportion of cases 1 in actual series is not significantly greater than that in simulates series. As we have already argued in Chap. 5, for a technical pattern to perform better in real series than in simulated series, significant heterogeneity should be observed in the proportions that correspond to the type of price behaviour realised after the pattern's formation. A heterogeneity though, that is attributed to the significant greater proportions of reaching the PT in real series than in simulated series $\left(f_1 > \widetilde{f}_1\right)$, along with significant lower proportions, $\left(f_3 < \widetilde{f}_3\right)$, that correspond to case3; the case where prices do not behave as TA asserts and a stop loss condition is activated. This type of superiority is observed only when RB are adopted on the NYSE components (Table 6.4).

[6] However, it should be noted that transaction costs were not considered in this study. Including transaction costs would exacerbated further the overall results.

Table 6.4 *z*-tests for comparing proportions and x^2-test for homogeneity

Case (i)	$f_i(\%)$	$\tilde{f}_i(\%)$	z-value	p-value	x^2	p-value
Panel A: NYSE, RB	$n_p = 8{,}792$, $\tilde{n}_p = 118{,}781$					
1	15.45	13.97	3.85	**0.0001**	83.24	**0.0000**
2	6.21	4.40	7.90	**0.0000**		
3	78.34	81.64	−7.66	**0.0000**		
Panel B: NYSE, RT	$n_p = 6{,}833$, $\tilde{n}_p = 51{,}260$					
1	7.05	5.42	5.53	**0.0000**	137.70	**0.0000**
2	6.59	10.80	−10.77	**0.0000**		
3	86.36	83.78	5.48	1.0000		
Panel C: NASDAQ, RB	$n_p = 4{,}632$, $\tilde{n}_p = 63{,}942$					
1	14.21	13.66	1.04	0.1503	8.18	**0.0167**
2	3.93	3.23	2.58	**0.0098**		
3	81.87	83.11	−2.17	**0.0149**		
Panel D: NASDAQ, RT	$n_p = 2{,}660$, $\tilde{n}_p = 22{,}350$					
1	5.83	4.43	3.27	**0.0005**	77.45	**0.0000**
2	7.33	13.02	−8.41	**0.0000**		
3	86.84	82.55	5.57	1.0000		

$f_i(\%)$ relative frequency of each case, n_p number of patterns identified, tildes indicate results obtained from simulated series, x^2-tests correspond to 2 degrees of freedom. p-values lower than 0.05 are in bold

6.4 Conclusions

In this chapter, an algorithmic methodology for identifying the RB and RT patterns is presented. This method was developed in order to capture the circular characteristics of these patterns. It is based on theoretical descriptions found in the literature regarding the identification of the pattern via visual assessment. Subsequently an analysis similar with the one presented in Chap. 5 for the zigzag patterns was conducted in order to assess the performance of the pattern. Findings are similar with the ones presented for the zigzag patterns. More precisely, RB and RT patterns are also identified successfully in randomly generated series. The price evolution after the pattern's formation was also broken down into three mutually exclusive cases and the relative frequencies of each case were calculated when real and simulated series were used. Results showed that the corresponding frequency counts are not distributed identically across real and simulated datasets. However, the source of this heterogeneity was only meaningful when the RB was applied on NYSE components, implying an important superiority in the patterns performance on real series against its performance on simulated series.

The presented RB identification algorithm was initially proposed in Zapranis and Tsinaslanidis (2012). In this study, we have also examined the profitability of this pattern, on a smaller dataset consisting of seven US tech stocks, using various fixed holding periods. We have shown that the pattern was outperformed by a simple trading range break out rule, and that there were more cases of significant excess returns when short term holding periods were used and earlier periods of the dataset used were considered.

Appendix 1: RBottoms Function

1.1. The Function

$$[Descriptives, Circles, Signals] = RBottoms(ys, Bounds, tWidth, tFit, w, pflag)$$

1.2. Description

The *RBottoms*$(\cdot)$ identifies RB patterns for a given price series. $RW(\cdot)$ is embedded in this function. Additionally, *RBcirlces*$(\cdot)$ is also embedded (see Appendix 1.3.1) and it performs all necessary calculations on the scaled data for defining the three homocentric circles and thus assess the possibility of a RB existence.

Inputs

1. *ys*: Price series (column vector).
2. *Bounds*: The bound rate used to construct the boundaries in where the majority of prices should fluctuate, in order to identify a RB pattern successfully.
3. *tWidth*: Threshold for the minimum accepted pattern's width (e.g. 15 days).
4. *tFit*: Threshold for the minimum accepted fit value (e.g. 0.9).
5. *w*: Is used to define the size of the rolling window which is 2w + 1 (see $RW(\cdot)$ in Appendix 1 of Chap. 2)
6. *pflag*: If 1 the function returns two plots illustrating the pattern identified with the best fit value.

Output

1. *Descriptives*: Structure array containing descriptive information of the n identified RB patterns. Particularly,
   - *Descriptives.CUPpnts_x*: Is a $n \times 3$ matrix. Columns contain the x-coordinates of the three characteristic points (P_1,P_2 and T_1) for each identified pattern.
   - *Descriptives.CUPpnts_y*: Is a $n \times 3$ matrix. Columns contain the corresponding y-coordinates.
   - *Descriptives.Fit*: A $n \times 1$ column vector containing patterns' fit values.
   - *Descriptives.Width*: A $n \times 1$ column vector containing patterns' widths.
   - *Descriptives.Depth*: A $n \times 1$ column vector containing patterns' depths.

2. *Circles*: Structure array containing information regarding the homocentric circles used for the RB recognition. Values are scaled, and they can be used to generate figures similar to Fig. 6.4.
   - *Circles.Ks*: Is a $n \times 2$ matrix containing coordinates of the circles' centers.
   - *Circles.Rs*: Is a $n \times 1$ matrix containing circles' radiuses.
   - *Circles.CrclX*: Is a $n \times 1$ cell array. Every cell contains a row vector with the x-coordinates used in the creation of the three homocentric circles.
   - *Circles.CrclY*: Is a $n \times 1$ cell array. Every cell contains a matrix of three rows with the corresponding y-coordinates.
   - *Circles.x_scaled*: Is a $n \times 1$ cell array. Identical to *Circles.CrclX*.
   - *Circles.y_scaled*: Is a $n \times 1$ cell array. Scaled values of the price series consisting the pattern.
3. *Signals*: Structure array containing buying signals under to perspectives. If l is the size of the examined price series (number of observations) then,
   - *Signals.Signals*1: Is a $l \times 1$ binary column vector containing ones at days where bullish trend is implied by the pattern and zeros elsewhere. In this case buy signal occur firstly at points P_2 and last for the m subsequent days considering that $m - 1$ is the pattern's width.
   - *Signals.Signals*2: In this case, only days at points P_2 indicate a buy signal.

1.3. Code

```
function [Descriptives,Circles,Signals]=...
RBottoms(ys,Bounds,tWidth,tFit,w,pflag)
%% Finding Critical Points
l=length(ys);
[Peaks,~]=RW(ys,w,0);
p=size(Peaks,1);% Number of Peaks
P_Indx=Peaks(:,2);% Index of Peaks
RBWidth=nan(p,1);
for i=1:p
  if isempty(find(ys(P_Indx(i,1)+1:l,1)>=...
      ys(P_Indx(i,1),1),1,'first'))==0
  RBWidth(i,1)=find(ys(P_Indx(i,1)+1:l,1)>=...
    ys(P_Indx(i,1),1),1,'first');
  end
end
RBP2s=P_Indx+RBWidth;% 3rd critical point
RBcrit_pnts=[P_Indx RBP2s nan(p,1)];% Critical Points
for i=1:p
  if isnan(RBcrit_pnts(i,2))==0
```

```
[~,RBcrit_pnts(i,3)]=min(ys(RBcrit_pnts(i,1):RBcrit_pnts(i,2)));
  end
end
% Eliminate nans and the cases where the pattern's width is lower
% than tWidth
RBcrit_pnts(RBWidth+1<=tWidth,2)=nan;
RBCUPpnts_x=RBcrit_pnts(isnan(RBcrit_pnts(:,2))==0,:);
RBCUPpnts_x(:,3)=RBCUPpnts_x(:,1)+RBCUPpnts_x(:,3)-1;
% x-coordinates for the three potential cup points
%(P1,P2,T1 respectively)
RBm=size(RBCUPpnts_x,1);
RBCUPpnts_y=ys(RBCUPpnts_x);% y-coordinates
%% Identification
% Preallocations:
RBactfit=zeros(RBm,1);% Actual Fit value
RBUform=zeros(RBm,1);% U formation
Circles.Ks=zeros(RBm,2);% Centers
Circles.Rs=zeros(RBm,1);% Radius
Circles.CrclX=cell(RBm,1);% x-coordinates for circles
Circles.CrclY=cell(RBm,1);% y-coordinates for circles
Circles.x_scaled=cell(RBm,1);% x-coordinates of data
Circles.y_scaled=cell(RBm,1);% y-coordinates of data
for i=1:RBm
  P1=[RBCUPpnts_x(i,1),RBCUPpnts_y(i,1)];
  P2=[RBCUPpnts_x(i,2),RBCUPpnts_y(i,2)];
  T1=[RBCUPpnts_x(i,3),RBCUPpnts_y(i,3)];
  % Scaling the data:(Focusing on the specific subparts of
  %the examined price series).
  w1=P2(1,1)-P1(1,1)+1;%width
  y_scaled=mapminmax(ys(P1(1,1):P2(1,1),1)',1,w1/2);
  x_scaled=P1(1,1):P2(1,1);

  Points.p1=[P1(1,1),y_scaled(1,1)];
  Points.p2=[P2(1,1),y_scaled(1,end)];
  Points.t1=[T1(1,1),1];% Scaled Y-coordinate of T is always 1.
[CrclX,CrclY,K,R]=RBcircles(Points.p1,Points.p2,Points.t1,
Bounds)
  % Keep Data for Circles
  Circles.x_scaled{i,1}=x_scaled;
  Circles.y_scaled{i,1}=y_scaled;
  Circles.CrclX{i,1}=CrclX;Circles.CrclY{i,1}=CrclY;
  Circles.Ks(i,:)=K(1,:);Circles.Rs(i,:)=R;

  Checking=zeros(1,P2(1,1)-P1(1,1)+1);
  for q=1:P2(1,1)-P1(1,1)+1
```

```
      if y_scaled(1,q)<=CrclY(1,q) && y_scaled(1,q)>=CrclY(3,q)
        Checking(1,q)=1;
      end
    end
    if sum(Checking)/length(Checking)>=tFit
      RBUform(i,1)=1;
    end
    RBactfit(i,1)=sum(Checking)/length(Checking);
end
% keep only the U patterns satisfy the tFit threshold
U_idx=find(RBUform);
RBCUPpnts_x=RBCUPpnts_x(U_idx,:);RBCUPpnts_y=RBCUPpnts_y
(U_idx,:)
Circles.Ks=Circles.Ks(U_idx,:);Circles.Rs=Circles.Rs(U_idx,:);
NotU=ismember(1:RBm,U_idx)==0;
Circles.x_scaled(NotU)=[];Circles.y_scaled(NotU)=[];
Circles.CrclX(NotU)=[];Circles.CrclY(NotU)=[];
Descriptives.CUPpnts_x=RBCUPpnts_x;
Descriptives.CUPpnts_y=RBCUPpnts_y;
Descriptives.Fit=RBactfit(U_idx,:);
Descriptives.Width=RBCUPpnts_x(:,2)-RBCUPpnts_x(:,1)+1;
Descriptives.Depth=...
    (max([RBCUPpnts_y(:,2),RBCUPpnts_y(:,1)],[],2)...
      -RBCUPpnts_y(:,3))./RBCUPpnts_y(:,3);
%% Plotting
if pflag==1 && isempty(Descriptives.CUPpnts_x)==0
[~,BUForm]=max(Descriptives.Fit);% Best Formation
figure(1)
plot(Circles.x_scaled{BUForm,1}',Circles.y_scaled{BUForm,1}')
hold on
plot(repmat(Circles.CrclX{BUForm,1}',1,3),...
Circles.CrclY{BUForm,1}','r'), hold off
figure(2)
plot(ys),hold on
plot(Descriptives.CUPpnts_x(BUForm,:),...
Descriptives.CUPpnts_y(BUForm,:),'r*'),hold off
end
%% Create trading signals
thesignals=zeros(1,1);
n=size(RBCUPpnts_x,1);
if n>0
    for i=1:n;
      if RBCUPpnts_x(i,2)+RBCUPpnts_x(i,2)-RBCUPpnts_x(i,1)<=l
      thesignals(RBCUPpnts_x(i,2):RBCUPpnts_x(i,2)+...
```

```
      RBCUPpnts_x(i,2)-RBCUPpnts_x(i,1),1)=1;
    else thesignals(RBCUPpnts_x(i,2):end)=1;
    end
  end
else
   thesignals=nan; % No Pattern Found
end
thesignals2=zeros(1,1);
if isempty(RBCUPpnts_x)==0
  thesignals2(RBCUPpnts_x(:,2),1)=1;% Buy signals
else
  thesignals2=nan;
end
Signals.Signals1=thesignals;
Signals.Signals2=thesignals2;
end
```

1.3.1. RBcircles Sub-function

```
function [CrclX,CrclY,K,R]=RBcircles(P1,P2,T1,Bounds)
% Calculation of the vertices (coordinates) of the isosceles
% triangle.
Ay=(P1(1,2)+P2(1,2))/2;
A=[P1(1,1),Ay]; C=[P2(1,1),Ay];
B=[(P1(1,1)+P2(1,1))/2,T1(1,2)];
Ax=A(1,1);Ay=A(1,2);
Bx=B(1,1);By=B(1,2);
Cx=C(1,1);Cy=C(1,2);
% R=a/2sin(A)
% D=sqrt(dx^2+dy^2)
by=A(1,2);%b(x,y) middle of A and C (bx=B(1,1));
a=sqrt((Bx-Cx)^2+(By-Cy)^2);
imitono=(by-By)/sqrt((Ax-Bx)^2+(Ay-By)^2);
R=a/(2*imitono);% Radious of the circle
Kx=B(1,1);Ky=By+R;
K=[Kx,Ky];% Center of the circle
CrclX=Ax:Cx;
% Y^2-2KyY+Ky^2-R^2+(X-Kx)^2=0
% aX^2+bX+c=0
aaa=1;% (for all Xs)
bbb=-2*Ky;% (for all Xs)
ccc=Ky^2-R^2+(CrclX-Kx*(ones(1,length(CrclX)))).^2;
DDD=bbb^2-4*aaa*ccc;% Determinant
```

```
Ys=[(-bbb+DDD.^(1/2))./(2*aaa);(-bbb-DDD.^(1/2))./(2*aaa)];
Ys(:,DDD<0)=Ky;
Yskept=min(Ys);
Rmax=R*(1+Bounds);Rmin=R*(1-Bounds);
% Calculating Circles 2 and 3.
% The center (Radious) is (is not) the same.
% aX^2+bX+c=0 Only ccc changes, so:
cccmin=Ky^2-Rmin^2+(CrclX-Kx*(ones(1,length(CrclX)))).^2;
cccmax=Ky^2-Rmax^2+(CrclX-Kx*(ones(1,length(CrclX)))).^2;
DDDmin=bbb^2-4*aaa*cccmin;DDDmax=bbb^2-4*aaa*cccmax;
Ysmin=[(-bbb+DDDmin.^(1/2))./(2*aaa);...
(-bbb-DDDmin.^(1/2))./(2*aaa)];
Ysmax=[(-bbb+DDDmax.^(1/2))./(2*aaa);...
(-bbb-DDDmax.^(1/2))./(2*aaa)];
Ysmin(:,DDDmin<0)=Ky;Ysmax(:,DDDmax<0)=Ky;
Yskeptmin=min(Ysmin);Yskeptmax=min(Ysmax);
CrclY=[Yskeptmin;Yskept;Yskeptmax];
end
```

References

Achelis SB (1995) Technical Analysis from A to Z. Probus Publishing, Chicago

Bo L, Linyan S, Mweene R (2005) Empirical study of trading rule discovery in China stock market. Expert Syst Appl 28:531–535

Duda R, Hart P (1973) Pattern classification and scene analysis. Wiley, New York

Edwards RD, Magee J (1997) Technical analysis of stock trends, 7th edn. John Magee, Inc., Boston

Leigh W, Frohlich CJ, Hornik S, Purvis R, Roberts TL (2008) Trading with a stock chart heuristic. IEEE Trans Syst Man Cybernet A Syst Hum 38(1):93–104

Leigh W, Hightower R, Modani N (2005) Forecasting the New York stock exchange composite index with past price and interest rate on condition of volume spike. Expert Syst Appl 28:1–8

Leigh W, Modani N, Hightower R (2004) A computational implementation of stock charting: abrupt volume increase as signal for movement in New York Stock Exchange Composite Index. Decis Support Syst 37:515–530

Leigh W, Modani N, Purvis R, Roberts T (2002a) Stock market trading rule discovery using technical charting heuristics. Expert Syst Appl 23:155–159

Leigh W, Paz N, Purvis R (2002b) Market timing: a test of a charting heuristic. Econ Lett 77:55–63

Leigh W, Purvis R, Ragusa JM (2002c) Forecasting the NYSE composite index with technical analysis, pattern recognizer, neural network, and genetic algorithm: a case study in romantic decision support. Decis Support Syst 32:361–377

Murphy J (1986) Technical analysis of the future markets: a comprehensive guide to trading methods and applications. Prentice Hall, New York

Pring MJ (2002) Technical analysis explained: the successful investor's guide to spotting investment trends and turning points, 4th edn. McGraw-Hill Book Company, New York

Wang J, Chan S (2007) Stock market trading rule discovery using pattern recognition and technical analysis. Expert Syst Appl 33:304–315

Wang J, Chan S (2009) Trading rule discovery in the US stock market: An empirical study. Expert Syst Appl 36:5450–5455

Zapranis A, Tsinaslanidis PE (2012) A novel, rule-based technical pattern identification mechanism: identifying and evaluating saucers and resistant levels in the US stock market. Expert Syst Appl 39(7):6301–6308. doi:10.1016/j.eswa.2011.11.079

Chapter 7
Technical Indicators

7.1 Introduction

This chapter presents a bundle of well-known technical indicators along with the corresponding formulas for their calculation. In the graphical examples provided we use parameters' values that we believe are the most commonly used, as implied by the literature. In addition, we present the manner that these technical tools will be used to derive trading signals (long, short and neutral positions). With this chapter we complete the universe of technical tools we are about to use in the subsequent chapter for testing the efficacy of TA. Although the calculation of these technical indicators is a relative simple task to implement is any programming language, the reader can find the corresponding Matlab functions in the financial toolbox of Matlab 2014.

The rest of this chapter is organized as follows. Sect. 7.2 focuses on moving averages, Sect. 7.3 presents moving average crossovers, Sect. 7.4 describes the moving average convergence divergence, Sect. 7.5 presents the relative strength index, Sect. 7.6 presents the Bollinger bands, Sect. 7.7 presents the momentum, Sect. 7.8 presents the price rate-of-change and Sect. 7.9 presents the highest high and lowest low indicators.

7.2 Moving Averages

Moving Averages (MAs) are considered as one of the main and celebrated technical analysis tools. As their name implies, they are rolling averages of a price series given a particular time span w. Versions of MAs mainly differ in the manner they

P.E. Tsinaslanidis, A.D. Zapranis, *Technical Analysis for Algorithmic Pattern Recognition*, DOI 10.1007/978-3-319-23636-0_7

are calculated. We can argue that all MAs are weighted, meaning that observations inside the rolling window are being assigned with weights in order produce the moving average. Different weighted schemes result is different types of MAs. For instance equal, linear and exponential weighting result in simple, linearly weighted and exponential MAs. The choice of time span affects the sensitivity of the MAs to the underlying series. Smaller values of w generate MAs more sensitive to price changes. Larger values of w generate smoother ones. In the Context of TA, MA can be used to smooth a price series, identify current trends, set up support and resistance levels and signal trend reversals. They can be used individually, combined [MA crossovers, Triple crossovers (Murphy 1999)] or as components of other technical tools (e.g. MACD). Subsequently, we provide the description, formal expressions and the basic interpretation of three main types of MAs. More precisely Sect. 7.2.1 focuses on the *Simple Moving Average* (SMA), Sect. 7.2.2 presents the *Linearly Weighted Moving Average* (LWMA), whilst Sect. 7.2.3 presents the *Exponential Moving Average* (EMA). Other types of MAs we do not present in this section are triangular MAs, variable MAs (Achelis 1995) and Hull MAs (Gorgulho et al. 2011).

7.2.1 Simple Moving Average

Let $\{P_t\}\equiv\{P_1,P_2,\ldots P_\ell\}$ be the price series of length ℓ, with ℓ being the most recent observation, and w the width of a rolling window. The simple SMA is defined as,

$$SMA_{t|w} = \frac{P_t + P_{t-1} + \ldots + P_{t-w+1}}{w}, \quad \forall\, t \in [w, \ell]. \tag{7.1}$$

The choice of w depends on the horizon of interest. Small values are chosen for identifying short-term trends and larger values for long-term trends. As implied by (7.1), the greater the value of w the smoother the estimated SMA.

According to TA, when $SMA_{t|w} < P_t$, investors' current expectations are higher than their average expectations over the last w periods. In this case it is probable that the market is bullish. The reverse stands when $SMA_{t|w} > P_t$. Trading signals are generated when prices breach SMA (7.2). Upward penetrations signal long positions whereas downward penetrations signal short positions (Achelis 1995).

$$g_{t|SMA_{t|w}} = \begin{cases} 1 \textit{ if } P_t > SMA_{t|w} \;\&\; P_{t-1} < SMA_{t-1|w} \\ -1 \textit{ if } P_t < SMA_{t|w} \;\&\; P_{t-1} > SMA_{t-1|w} \\ 0 \textit{ otherwise} \end{cases} \tag{7.2}$$

One of the most common criticisms against SMA is that they generate late trading signals. This is attributed to the fact that for its calculation an equal weighting is assigned to all prices within the considered time span, w. Murphy (1999) states that

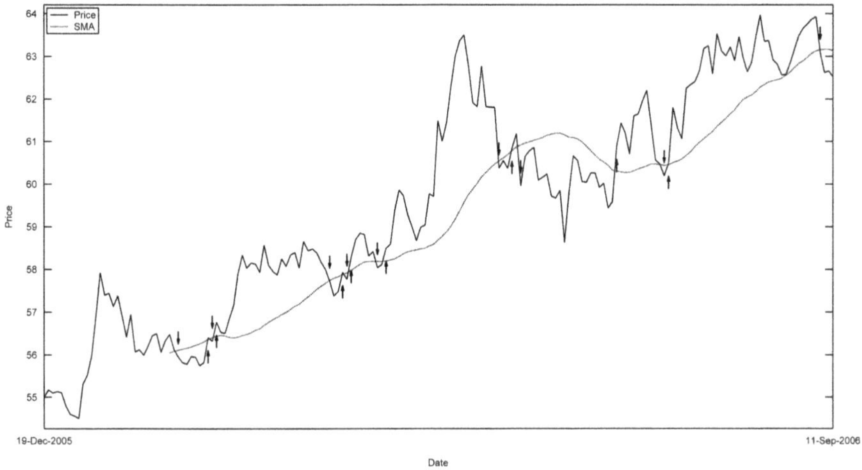

Fig. 7.1 Long (*upwards arrows*) and short (*downwards arrows*) positions implied by the 30-days SMA ($SMA_{t|30}$) on the HSBC Holdings plc traded in NYSE

some analysts believe that a heavier weighting should be attributed to more recent observations. Another criticism against its usefulness also mentioned in Murphy (1999) is the fact that it considers only prices within the rolling window. Finally, when prices breach a MA and then return to their average in a short notice, following trading signals of a MA might result in losses, due to high transaction costs. This is attributed to the fact that trading positions are being frequently opened and closed on the same price levels. Achelis refers to this phenomenon as "traders' remorse" particularly in cases where long term MAs are used. Figure 7.1 presents a 30-days SMA drawn on the HSBC price series. It is obvious that the first day the SMA is plotted is the 30th trading day. After that day, long and short trading signals occur according to the aforementioned description. In addition, as we already mentioned there are cases where long and short trading signals occur in a short time span on the same price level. More specifically, in a volatile period when sensitive indicators are used (like a short-term MA) long or short positions are reversed in a short time. In technical jargon this is called "whipsaw". The number of these cases can be reduced by adopting additional filters after each trading signal (see Chap. 1) or by using moving average crossovers discussed later in Sect. 7.3.

7.2.2 Linearly Weighted Moving Average

Obviously, when calculating a SMA, observations used are equally weighted. Alternative, linearly weighted MAs can be used. In this case, greater weights are assigned to more recent observations dealing with the weighting "problem" mentioned in Murphy (1999). For its calculation, each observation is multiplied by its position inside the rolling window.

$$LWMA_{t|w} = \frac{wP_t + (w-1)P_{t-1} + \ldots + 2P_{t-w+2} + P_{t-w+1}}{w(w+1)/2}, \quad \forall\, t \in [w, \ell] \quad (7.3)$$

The denominator of (7.3) is a triangular number and it represents the sum of the weights assigned on each observation (i.e. $w(w+1)/2 = 1 + 2 + \ldots w$). Again, as indicated in (7.3), the wth day is the first day that a w-day LWMA can be a calculated. The interpretation, of this type of MA is similar to that of SMA and thus the reader may refer back to (7.2).

7.2.3 *Exponential Moving Average*

The EMA can be calculated by the following recurrent formula (Achelis 1995)

$$EMA_{t|w,\lambda} = (1-\lambda)EMA_{t-1|w,\lambda} + \lambda P_t, \quad \forall\, t \in [w+1, \ell], \qquad (7.4)$$

where $0 \leq \lambda \leq 1$, is the multiplier, and we set the starting value $EMA_{w|w,\lambda} = SMA_{w|w}$. It is a common practice in TA to set $\lambda = \lambda^* \equiv 2/(w+1)$. However, a technician may use any value for λ depending on the importance he wants to give to more recent observations. The greater the value of λ, the greater the importance of current observation in determining the new point of EMA. This last type of MA deals with both weighting issues associated with the SMA: the weighting problem and the problem of not considering all available prices in the calculation (Murphy 1999).

7.3 Moving Averages Crossovers

Two MAs of different time spans w_S, w_L, where $w_S < w_L$, can also be used in combination to retrieve trading signals.[1] Long and short positions are signaled when the short term MA breaches the longer term MA upwards and downwards respectively. Quantitatively, *moving average crossovers* (MAC) can be calculated by (7.5), although other types of MAs (like EMAs) can be used alternatively. Finally, (7.6) shows how to derive the corresponding trading signals.

[1] Murphy (1999) refers to this technique as a double crossover method to distinguished it from the triple crossover method whereby three MAs are used instead of two.

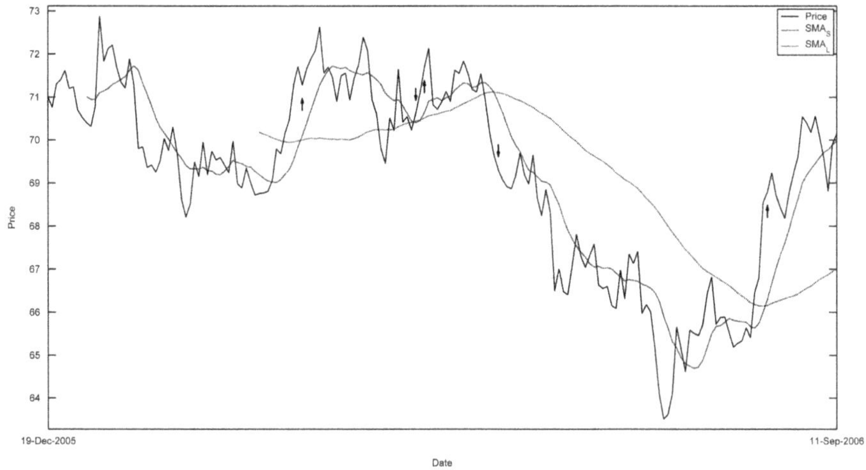

Fig. 7.2 Long (*upwards arrows*) and short (*downwards arrows*) positions implied by the $MAC_{t|10,50}$ on the IBM traded in NYSE

$$MAC_{t|w_S,w_L} = SMA_{t|w_S} - SMA_{t|w_L}, \quad \forall\, t \in [w_L, \ell] \tag{7.5}$$

$$g_{t|MAC_{t|w_S,w_L}} = \begin{cases} 1 \; if \; MAC_{t|w_S,w_L} > 0 \;\&\; MAC_{t-1|w_S,w_L} < 0 \\ -1 \;\; if \; MAC_{t|w_S,w_L} < 0 \;\&\; MAC_{t-1|w_S,w_L} > 0 \\ 0 \; otherwise \end{cases} \tag{7.6}$$

Figure 7.2 displays two SMAs drawn on the International Business Machines Corporation (IBM) stock traded at NYSE. In this case 10-days and 50-days SMAs are used, i.e. $w_S = 10$ and $w_L = 50$. It can be argued that when using a single MA, say $SMA_{t|50}$ is an equivalent process of using $MAC_{t|1,50}$. That is because $SMA_{t|1} \equiv P_t$ and in both cases a $SMA_{t|50}$ is adopted to derive trading signals. However, when using MAC, increasing the value of w_S smoothens the initial price series on which the long-term MA is applied. Thus the number of whipsaws is reduced but signals produced by using MAC lag the market a bit more than those produced with a single MA (Murphy 1999).

7.4 Moving Average Convergence Divergence

Moving Average Convergence Divergence (MACD) is plotted above the price chart, and for its interpretation the signal line is also required. MACD is defined as the difference between two different EMAs of periods w_S, w_L, where $w_S < w_L$ (7.7). The signal line (SL) is a w_{sig}-period EMA applied on the MACD (7.8). In the context of TA, technicians usually set $w_L = 26$, $w_S = 12$ and $w_{signal} = 9$ days (Murphy 1999). Formally this is,

$$MACD_{t|w_S,w_L,\lambda} = EMA_{t|w_S,\lambda} - EMA_{t|w_L,\lambda}, \quad \forall\, t \in [w_L, \ell], \tag{7.7}$$

and the signal line is

$$SL_{t|w_{sig},\lambda} = EMA^{\{MACD_t\}}_{t|w_{sig},\lambda}, \quad \forall\, t \in [w_L + w_{sig} - 1, \ell]. \tag{7.8}$$

In (7.8), the superscript $\{MACD_t\}$ indicates that the EMA is applied on the MACD series instead of the original price series.

The trading rule we adopt produces long signals when the MACD crosses upwards the signal line and short signals when the reverse stands (7.9).

$$g_{t|MACD_{t|w_S,w_L,\lambda},SL_{t|w_{sig},\lambda}} = \begin{cases} 1 \text{ if } MACD_{t|w_S,w_L,\lambda} > SL_{t|w_{sig},\lambda} \;\&\; MACD_{t-1|w_S,w_L,\lambda} < SL_{t-1|w_{sig},\lambda} \\ -1 \text{ if } MACD_{t|w_S,w_L,\lambda} < SL_{t|w_{sig},\lambda} \;\&\; MACD_{t-1|w_S,w_L,\lambda} > SL_{t|w_{sig},\lambda} \\ 0 \text{ otherwise} \end{cases} \tag{7.9}$$

However, as it happens with many technical tools its interpretation varies. For example, Chong and Ng (2008) derive buy signals when MACD crosses the zero line from below, while sell signals are produced when MACD crosses the zero line from above. Rosillo et al. (2013)) use MACD in a different manner: buy recommendations are generated when both the $MACD_{t|w_S,w_L,\lambda} < SL_{t|w_{sig},\lambda} < 0$ while sell recommendations are triggered when $MACD_{t|w_S,w_L,\lambda} > SL_{t|w_{sig},\lambda} > 0$. The chart in Fig. 7.3 shows the MACD and SL (upper graph) derived from the McDonald's Corp. (lower graph) traded at NYSE. Again, buy and sell recommendations are symbolized with up and down arrows.

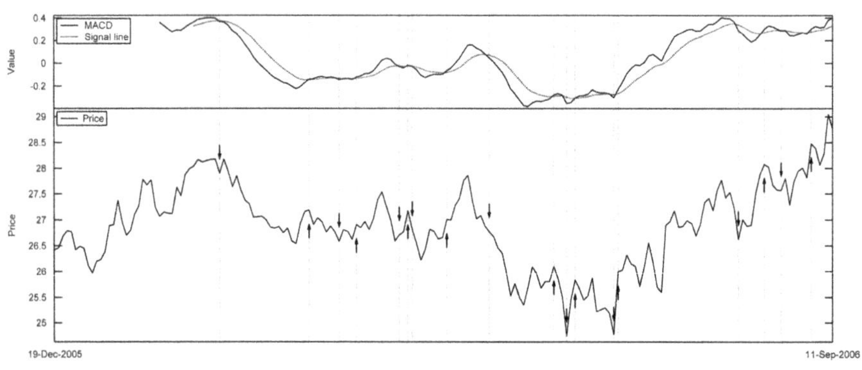

Fig. 7.3 Long (*upwards arrows*) and short (*downwards arrows*) positions implied by the $MACD_{t|12,26\lambda^*}$ and $SL_{t|9,\lambda^*}$ on the McDonald's Corp. traded in NYSE. For ease of viewing, vertical lines are used to locate the trade recommendations drawn on the price chart on the MACD-SL graph

7.5 Relative Strength Index

Relative Strength Index (RSI) is a technical oscillator that ranges between 0 and 100. This technical tool was designed by J. Welles Wilder Jr and is explained in detail in (Wilder 1978). Several steps are needed for its calculation. First, on time $t \in [2, \ell]$, (7.10) and (7.11) define upward and downward changes as ΔP_t^+ and ΔP_t^- respectively.

$$\Delta P_t^+ = \max(P_t - P_{t-1}, 0) \tag{7.10}$$

$$\Delta P_t^- = |\min(P_t - P_{t-1}, 0)| \tag{7.11}$$

Then it is required by the user to define a time period to calculate the RSI. Using a 14-day RSI is a popular choice by technicians. However, before calculating RSI, the *Relative Strength* (RS) at time t is defined as

$$RS_{t|w} = \frac{\sum_{i=t-w+1}^{t} \Delta P_i^+}{\sum_{i=t-w+1}^{t} \Delta P_i^-}, \quad \forall\, t \in [w + 1, \ell]. \tag{7.12}$$

If $RS_{i|w}$ is not defined (i.e. the denominator of (7.12) is zero), then RSI takes the maximum value of 100. Otherwise, we use its basic formula:

$$RSI_{t|w} = \begin{cases} 100, \ \mathit{if} \sum_{i=t-w+1}^{t} \Delta P_i^- = 0 \\ 100 - \dfrac{100}{1 + RS_{t|w}}, \quad \mathit{otherwise} \end{cases} \tag{7.13}$$

It is common practice to draw two levels; one upper (*ul*) and one lower (*dl*) level, which are equally distanced from the 50-level, say $ul = 70$ and $dl = 30$. Again there are various ways technicians interpret RSI. We will adopt the same trading rule with the one described in (Wilder 1978; Henderson 2002; Rosillo et al. 2013). Purchase recommendations occur when RSI crosses the lower level from below, while sale recommendations are triggered when RSI crosses the upper level from above i.e.,

$$g_{t|RSI_{t|w}, ul, dl} = \begin{cases} 1 \ \mathit{if} \ RSI_{t|w} > dl \ \& \ RSI_{t-1|w} < dl \\ -1 \ \mathit{if} \ RSI_{t|w} < ul \ \& \ RSI_{t-1|w} > ul \, . \\ 0 \ \mathit{otherwise} \end{cases} \tag{7.14}$$

For alternative and further uses of RSI the reader can see (Achelis 1995; Murphy 1999; Pring 2002; Chong and Ng 2008). Figure 7.4 shows an example where the 14-day RSI is used, with $ul = 70$ and $dl = 30$, to derive trading recommendations for the Procter & Gamble company.

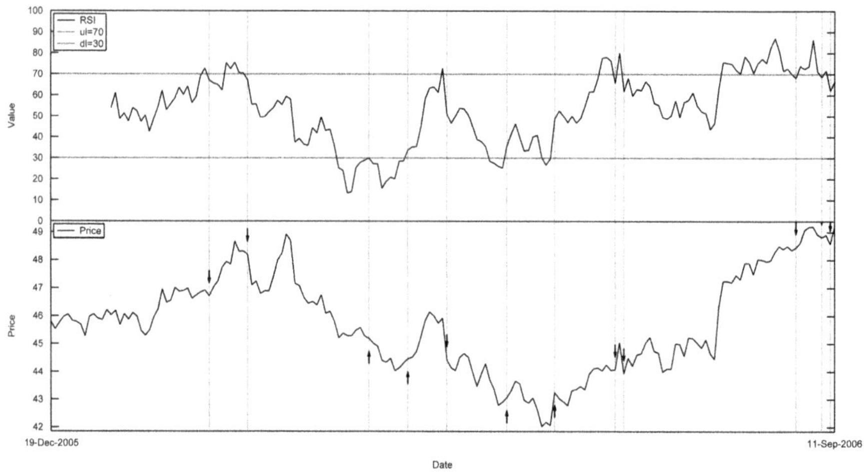

Fig. 7.4 Long (*upwards arrows*) and short (*downwards arrows*) positions implied by the $RSI_{t|14}$ on the Procter & Gamble traded in NYSE. For ease of viewing, *vertical lines* are used to locate the trade recommendations drawn on the price chart on the RSI graph

7.6 Bollinger Bands

Bollinger Bands (BB) is another celebrated technical tool, widely adopted by technicians. They were initially introduced by John Bollinger in the 1980s, and they are drawn usually on the price chart. They consist of three bands: The middle one (BB^{mid}) is a MA which is used to smooth the price series and observe the trend. Although various types of MAs can be used for this purpose a common practice is to use a SMA (Achelis 1995; Lim et al. 2013). The upper (BB^{up}) and lower (BB^{low}) bands are calculated by adding $\pm k$ standard deviations (σ) to BB^{mid} (7.15). The common length of time span used is 20 days.

$$BB_{t|w,k} = \begin{cases} BB^{mid}_{t|w} = SMA_{t|w} \\ BB^{up}_{t|w,k} = BB^{mid}_{t|w} + k\sigma_{t|w}, \quad \forall\, t \in [w, \ell] \\ BB^{low}_{t|w,k} = BB^{mid}_{t|w} - k\sigma_{t|w} \end{cases} \tag{7.15}$$

In (7.15) $\sigma_{t|w}$ is defined as:

$$\sigma_{t|w} = \sqrt{\frac{\sum_{i=t-w+1}^{t} \left(P_i - BB^{mid}_{i|w}\right)^2}{w}} \tag{7.16}$$

Buy considering σ in the upper and lower bands calculation, BB will widen in volatile periods and narrow in periods of low volatility. The trading rule we adopt generates buy recommendations when prices move outside the upper bands or when price crossed the lower band from below (Leung and Chong 2003; Lim et al. 2013). Reversely, when prices move outside the lower bands or when price crosses the upper band from above, sale recommendations will be triggered (7.17).

$$g_{t|BB_{t|w,k}} = \begin{cases} 1 \; if \; P_t > BB^{up}_{t|w,k} \; \& \; P_{t-1} > BB^{up}_{t-1|w,k} \\ 1 \; if \; P_t > BB^{low}_{t|w,k} \; \& \; P_{t-1} < BB^{low}_{t-1|w,k} \\ -1 \; if \; P_t < BB^{low}_{t|w,k} \; \& \; P_{t-1} < BB^{low}_{t-1|w,k} \\ -1 \; if \; P_t < BB^{up}_{t|w,k} \; \& \; P_{t-1} > BB^{up}_{t-1|w,k} \\ 0 \; otherwise \end{cases} \tag{7.17}$$

Figure 7.5 shows long and short recommendations generated by the adopted trading rule (7.17) on the AT&T Inc. traded at NYSE. As illustrated, when prices move outside the external bands, the rule might generate subsequent trading recommendations of the same type. This can be dealt, with a cost of realizing trading signals lagging the market, by adjusting (7.17) and requiring more consecutive days where price moves outside the bands.

Of course, these are not the only means by which BBs can be interpreted. Murphy (1999) suggests using the BB^{up} (BB^{low}) as upper (lower) price targets, when prices cross from below (above) the BB^{mid}. Further suggestions for interpreting BB are also presented in Achelis (1995).

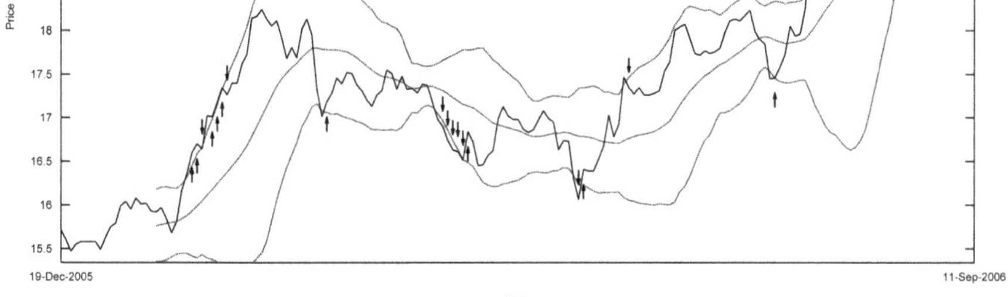

Fig. 7.5 Long (*upwards arrows*) and short (*downwards arrows*) positions implied by the $BB_{t|20,2}$ on the AT&T Inc. traded in NYSE

7.7 Momentum

The *momentum* indicator (MOM) is usually plotted above the price chart (like RSI) and measures a security's price change on a rolling basis for a given time span, w (7.18).

$$MOM_{t|w} = P_t - P_{t-w}, \quad \forall\, t \in [w+1, \ell] \tag{7.18}$$

Of course different values of w can be used, however setting w with 12 is a common choice among technicians (Rosillo et al. 2013). One way to interpret MOM and derive trading recommendations is by using a MA on it, in a manner similar to the one used in MACD. In particular, (7.19) shows the calculation of this MA that we also call signal line.

$$SL_{t|w_{sig}} = SMA_{t|w_{sig}}^{\{MOM_t\}}, \quad \forall\, t \in \left[w + w_{sig} - 1, \ell\right]. \tag{7.19}$$

Again, the superscript $\{MOM_t\}$ indicates that the SMA is applied on the MOM series instead of the original price series. Buy and sale recommendations occur when MOM crosses from below and from above the MOM respectively (7.20).

$$g_{t|MOM_{t|w}, SL_{t|w_{sig}}} = \begin{cases} 1 \; if \; MOM_{t|w} > SL_{t|w_{sig}} \; \& \; MOM_{t-1|w} < SL_{t-1|w_{sig}} \\ -1 \; if \; MOM_{t|w} < SL_{t|w_{sig}} \; \& \; MOM_{t-1|w} > SL_{t-1|w_{sig}} \\ 0 \; otherwise \end{cases} \tag{7.20}$$

This method is suggested in (Achelis 1995) and is included in Pring's (2002) descriptions, although other ways of interpretation can be found in the literature as well. For example (Murphy 1999; Rosillo et al. 2013) mention that trade signals occur when MOM crosses the zero level. Figure 7.6 presents trading signals

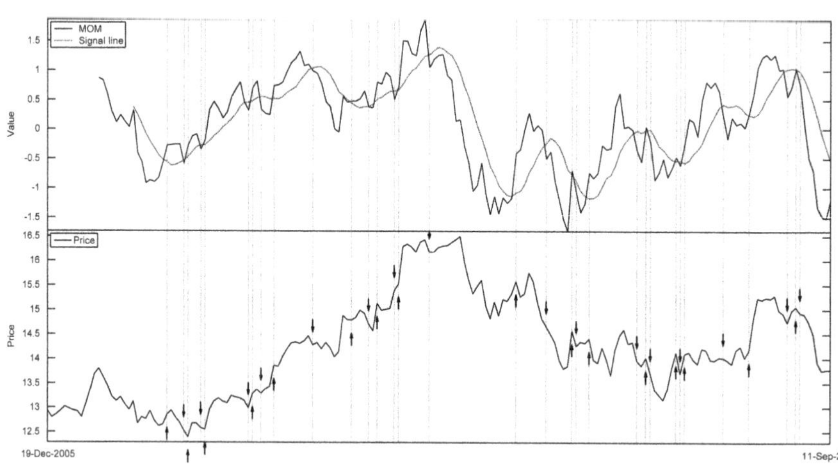

Fig. 7.6 Long (*upwards arrows*) and short (*downwards arrows*) positions implied by the $MOM_{t|12}$ on the Nokia Corp. traded in NYSE

generated by (7.20) when MOM was adopted on the Nokia stock price series, with $w = 12$ and $w_{sig} = 9$.

7.8 Price Rate-of-Change

The price *Rate-of-Change* (ROC) indicator displays a security's price change on a rolling basis for a given time span, w, like MOM, with the difference that this price change is expressed as a percentage (7.21). For short-term trading setting $w = 12$ is a popular choice for the time span used by technicians.

$$ROC_{t|w} = 100(P_t/P_{t-w} - 1), \quad \forall\, t \in [w+1, \ell] \tag{7.21}$$

Regarding the interpretation of price ROC, Achelis (1995) provides an example where trading signals are derived when the indicator leaves an oversold or overbought level. In particular, purchase recommendations are triggered when ROC crosses the oversold (lower) level from below, and sale signals occur when ROC crosses the overbought level from above. These levels can be drawn optically, and they should isolate historical local extrema on the security's ROC. Obviously, as Achelis mentions, these levels will depend on the security being analysed and overall market conditions. We propose using the $RW(\cdot)$ presented in Chap. 2, for identifying historical regional locals on the price ROC and then with a rolling average to draw the oversold and overbought levels (ul_t, dl_t). This process is subsequently described.

First we define an initial sample of size, w_{in} (say one trading year, $w_{in} = 250$).

Let $\{ROC_t\}_{t\in[t-w_{in}+1,t]}$ be the part of the price ROC series for the period $[t - w_{in} + 1, t]$. At time $t = w_{in}$, n regional peaks are identified by adopting the $RW(\cdot)$ on $\{ROC_t\}_{t\in[1,w_{in}]}$. The first upper level (ul_t) is defined at time $t = w_{in}$ as the mean of these extremes. Subsequently, the second upper level (ul_{t+1}) is defined at time $t = w_{in} + 1$ by using the same process and considering peaks realised in the $[w_{in} + 1 - w_{in} + 1, w_{in} + 1] = [2, w_{in} + 1]$. The above routine is repeated until $t = \ell$ and can be formally expressed as

$$ul\Big|_{t\,|\,w_{RW},\{ROC_t\}_{t\in[t-w_{in}+1,t]}} = \sum\nolimits_{i=1}^{n} Peaks_{i|\{ROC_t\}_{t\in[t-w_{in}+1,t]}}/n, \quad \forall\, t \in [w_{in}, \ell], \tag{7.22}$$

where w_{RW} length of the rolling window used for the identification of regional locals. Similarly, by using the m identified regional bottoms the oversold levels, dl_t, can be defined as

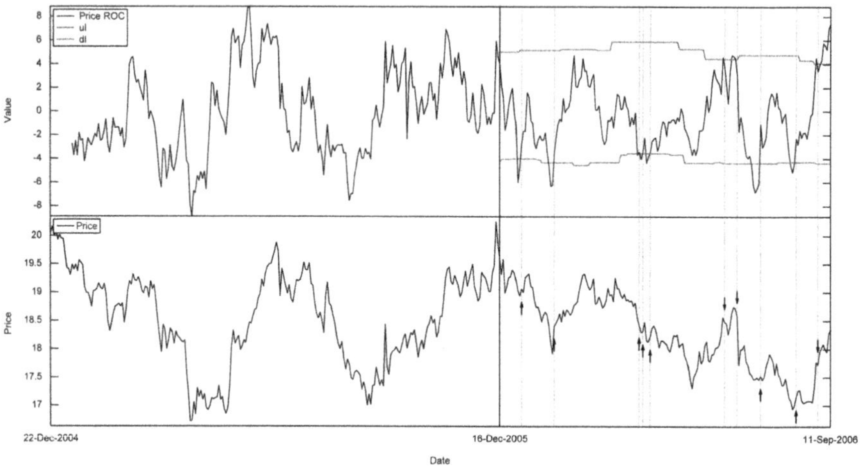

Fig. 7.7 Long (*upwards arrows*) and short (*downwards arrows*) positions implied by the $ROC_{t|12}$ on the Nike Inc. traded in NYSE

$$dl\Big|_{t|w_{RW},\{ROC_t\}_{t\in[t-w_{in}+1,t]}} = \sum_{i=1}^{m} Bottoms_{i|\{ROC_t\}_{t\in[t-w_{in}+1,t]}}/m, \quad \forall\, t \in [w_{in}, \ell]. \tag{7.23}$$

Using the above rolling procedure allows the user to derive time varying overbought and oversold levels. Thus, trading recommendations occur by using (7.24).

$$g_{t|ROC_{t|\cdot},ul_{t|\cdot},dl_{t|\cdot}} = \begin{cases} 1 \; if \; ROC_{t|w} > dl_{t|\cdot} \;\&\; ROC_{t-1|w} < dl_{t-1|\cdot} \\ -1 \;\; if \; ROC_{t|w} < ul_{t|\cdot} \;\&\; ROC_{t-1|w} > ul_{t-1|\cdot} \\ 0 \; otherwise \end{cases} \tag{7.24}$$

Figure 7.7 shows purchase and sale recommendations generated by adopting a ROC trading rule (7.24) on the NIKE inc. traded at NYSE for the period 16-Dec-2005 to 11-Sep-2006. Parameter values used in this illustration are $w = 12$, $w_{RW} = 12$ and $w_{in} = 250$. Note that the ROC series calculation begins at $t = w + 1 = 13$, a preceding period of 250 days is used to derive *ul* and *dl*, whilst trading signals are realised after this period.

7.9 Highest High and Lowest Low

A simple way to express support and resistance levels is by using the *Highest High* (HH) and *Lowest Low* (LL) indicator. A HH (LL) level at time *t* is defined as the maximum (minimum) price over the previous *w* observations [(7.25) and (7.26)].

$$HH_{t|w} = \max(P_{t-1}, P_{t-2}, \ldots, P_{t-w}), \quad \forall\, t \in [w+1, \ell] \tag{7.25}$$

$$LL_{t|w} = \min(P_{t-1}, P_{t-2}, \ldots, P_{t-w}), \quad \forall\, t \in [w+1, \ell] \tag{7.26}$$

Perhaps, due to its calculative simplicity this technical tool has been widely assessed in various scientific articles, like (Brock et al. 1992; Hudson et al. 1996; Sullivan et al. 1999; Marshall et al. 2008; De Zwart et al. 2009; Cheung et al. 2011). Regarding its interpretation, long positions are recommended when price crosses the HH from above, whilst short positions are signaled when price crosses the LL from below (7.27).

$$g_{t|HH_{t|w}, LL_{t|w}} = \begin{cases} 1 \; if \; P_t > HH_{t|w} \;\; \& \; P_{t-1} < HH_{t-1|w} \\ -1 \;\; if \; P_t < LL_{t|w} \;\; \& \; P_{t-1} > LL_{t-1|w} \\ 0 \; otherwise \end{cases} \tag{7.27}$$

References

Achelis SB (1995) Technical analysis from A to Z. Probus Publishing, Chicago

Brock W, Lakonishok J, Lebaron B (1992) Simple technical trading rules and the stochastic properties of stock returns. J Financ 48(5):1731–1764

Cheung W, Lam KSK, Yeung H (2011) Intertemporal profitability and the stability of technical analysis: evidences from the Hong Kong stock exchange. Appl Econ 43:1945–1963

Chong TT-L, Ng W-K (2008) Technical analysis and the London stock exchange: testing the MACD and RSI rules using the FT30. Appl Econ Lett 15:1111–1114

De Zwart G, Markwat T, Swinkels L, van Dijk D (2009) The economic value of fundamental and technical information in emerging currency markets. J Int Money Financ 28:581–604

Gorgulho A, Neves R, Horta N (2011) Applying a GA kernel on optimizing technical analysis rules for stock picking and portfolio composition. Expert Syst Appl 38(11):14072–14085

Henderson C (2002) Currency strategy: the practitioner's guide to currency investing, hedging and forecasting. Wiley, Hoboken, NJ

Hudson R, Dempsey M, Keasey K (1996) A note on weak form efficiency of capital markets: the application of simple technical trading rules to UK prices 1935–1994. J Bank Financ 20:1121–1132

Leung JM-J, Chong TT-L (2003) An empirical comparison of moving average envelopes and Bollinger Bands. Appl Econ Lett 10(6):339–341

Lim KJS, Hisarly TT, Ng SH (2013) The profitability of a combined signal approach: Bollinger bands and the ADX. Int Federation Techn Anal J 2014 Edition

Marshall BR, Cahan RH, Cahan JM (2008) Can commodity futures be profitably traded with quantitative market timing strategies? J Bank Financ 32:1810–1819

Murphy J (1999) Technical analysis of the financial markets: a comprehensive guide to trading methods and applications. New York Institute of Finance, New York

Pring MJ (2002) Technical analysis explained: the successful investor's guide to spotting investment trends and turning points, 4th edn. McGraw-Hill Book Company, New York

Rosillo R, De la Fuente D, Brugos JAL (2013) Technical analysis and the Spanish stock exchange: testing the RSI, MACD, momentum and stochastic rules using Spanish market companies. Appl Econ 45(12):1541–1550

Sullivan R, Timmermann A, White H (1999) Data-snooping, technical trading rule performance, and the bootstrap. J Financ LIV(5):1647–1691

Wilder JWJ (1978) New concepts in technical trading systems. Hunter Publishing Company, Greensboro, NC

Chapter 8
A Statistical Assessment

8.1 Introduction

This chapter examines the performance of TA, by implementing all technical patterns and indicators presented previous chapters, on 18 financial market indices around the world, for the requested period of 1998–2014. Parameters used for defining these trading rules are excerpted from the literature, and are the most commonly used. Following the pioneered methodology applied in Brock et al. (1992) ordinary statistical tests are presented first and bootstrap techniques are subsequently implemented. A first indication regarding the generalised efficacy of the technical tools considered is obtained by adopting ordinary statistical tests, which include *Student's t-tests* and *Bernoulli trials*. However, distributional assumptions made by these test are not likely to occur in actual price series. Real price series present leptokurtosis, autocorrelation and conditional heteroskedasticity. To deal with this problem a bootstrap methodology is followed where distributions are generated from the increments of simulated null models (GARCH-m and E-GARCH) after ensuring the independency and homoskedasticity of their increments.

The rest of this chapter is organized as follows. In Sect. 8.2 the dataset used, the technical tools considered and the holding period selection is presented. In Sect. 8.3 results obtained from an ordinary statistical assessment are presented, whilst Sect 8.4 presents results obtained from the bootstrap methodology. Finally, in Sect. 8.5 we conclude.

P.E. Tsinaslanidis, A.D. Zapranis, *Technical Analysis for Algorithmic Pattern Recognition*, DOI 10.1007/978-3-319-23636-0_8

8.2 Dataset, Technical Tools and the Choice of Holding Period

This section precedes the empirical analysis that this chapter presents, and describes the dataset used (Sect. 8.2.1), the technical tools considered along with their parameters' values (Sect. 8.2.2), and the design of holding periods (Sect. 8.2.3) that will be adopted after the realization of technical trading signals.

8.2.1 *Dataset*

The initial dataset considered in this chapter consists of 18 major world indices (Table 8.1). Adjusted daily closing prices for the period Jan-1998 until Dec-2012 were downloaded from Bloomberg database. We applied a filter similar with the one adopted in Lo et al. (2000), Marshall et al. (2009) and Zapranis and Tsinaslanidis (2012) and remaining missing values were filled with linear interpolation. On Table 8.1 a symbol idx_i is allocated to each market index which is going to be used, for brevity reasons, in subsequent tables and graphs for presenting generated results.

8.2.2 *The Universe of Technical Trading Strategies*

Table 8.2 presents the bundle of technical tools adopted to derive trading signals and asses their performance. The assembly of technical tools consists of ten technical indicators/oscillators (see Chap. 7) and eight technical patterns (see Chaps. 4, 5 and 6). The symbol x_i has also been allocated to each technical tool for referencing purposes. The choice of the parameters used for each x_i has been discussed in the corresponding chapters. Considering the technical indicators/

Table 8.1 Major world indices

idx_i	Index name	idx_i	Index name	idx_i	Index name
idx_1	DOW JONES (INDU)	idx_7	EURO Stoxx (SX5E)	idx_{13}	AEX (AEX)
idx_2	S&P 500 (SPX)	idx_8	FTSE 100 (UKX)	idx_{14}	OMX STKH30 (OMX)
idx_3	NASDAQ (CCMP)	idx_9	CAC 40 (CAC)	idx_{15}	SWISS MKT (SMI)
idx_4	TSX (SPTSX)	idx_{10}	DAX (DAX)	idx_{16}	NIKKEI (NKY)
idx_5	MEX IPC (MEXBOL)	idx_{11}	IBEX 35 (IBEX)	idx_{17}	HANG SENG (HSI)
idx_6	IBOVESPA (IBOV)	idx_{12}	FTSE MIB (FTSEMIB)	idx_{18}	ASX 200 (AS51)

The Bloomberg ticker for every index is presented in brackets

Table 8.2 Universe of adopted technical tools with their parameters' values

x_i	Technical tool	Parameter(s)
Panel A:	Technical indicators/oscillators	
x_1	$SMA_{t\|w}$ (Short term)	$w = 10$
x_2	$SMA_{t\|w}$ (Medium term)	$w = 50$
x_3	$SMA_{t\|w}$ (Long term)	$w = 200$
x_4	$MAC_{t\|w_S,w_L}$	$w_S = 10, w_L = 50$
x_5	$MACD_{t\|w_L,w_S,\lambda}$ and $SL_{t\|w_{sig},\lambda}$	$w_S = 12, w_L = 26,\ w_{sig} = 9,\ \lambda = \lambda^*$
x_6	$RSI_{t\|w}$	$w = 14,\ ul = 70,\ dl = 30$
x_7	$BB_{t\|w,k}$	$w = 20,\ k = 2$
x_8	$MOM_{t\|w}$ and $SL_{t\|w_{sig}}$	$w = 12, w_{sig} = 9$
x_9	$ROC_{t\|w}$	$w = 12, w_{train} = 250$
x_{10}	$HH_{t\|w}$ and $LL_{t\|w}$	$w = 50$
Panel B:	Technical patterns	
x_{11}	$HSAR(\cdot)$	$w = 25, x = 0.05$
x_{12}	$HS(\cdot)$	$w = 7$
x_{13}	$TTTB(\cdot)$	$w = 15$
x_{14}	$DTDB(\cdot)$	$w = 15$
x_{15}	$FLAGS(\cdot)$	$w = 3$
x_{16}	$PENNANTS(\cdot)$	$w = 2$
x_{17}	$WEDGES(\cdot)$	$w = 7$
x_{18}	$RBottoms(\cdot)$ and $RTops(\cdot)$	$Bounds = 0.3, tWidth = 15, tFit = 0.9, w = 10$

oscillators, parameters chosen were those conceived as most commonly used in TA literature.

These technical rules enclose bullish and bearish versions that generate buy and sell signals accordingly. For example, x_{12} signals long positions i.e., buy signals, when the inverse version of HS is identified whereas its normal form signals short positions, i.e. sell signals.

8.2.3 Holding Periods

An initial position, either long or short, is taken according with the descriptions provided in Chaps. 4,5, 6 and 7 for each technical tool x_i. Let $x_{i,t} \in \{-1, 0, 1\}$ be the trading signal produced from x_i at time t, which takes three values: 1, -1 and 0 for long, short and neutral positions respectively. The assessment of the TA's profitability presented in this chapter implements HP-day holding periods after opening a position. Ordinary each position closes after HP days. However, after the realization of a buy (sell) signal, if a sell (buy) signal occurs within the next HP days the initial position closes earlier. Quantitatively, the initial position closes at time

t either if $x_{i,t} = x_{i,t-1} = x_{i,t-2} = \cdots = x_{i,t-HP+1} = 0$ and $x_{i,t-HP} \in \{-1, 1\}$, or if $-x_{i,t} \in \{x_{i,t-1}, x_{i,t-2}, \cdots, x_{i,t-HP}\}$.

8.3 An Ordinary Statistical Assessment

In this section it is examined whether there is any statistical significance in the returns generated by the implemented universe of technical tools on 18 major world indices. If TA has predictive power on a price series, then these returns would differ considerably from the unconditional distributions of daily returns observed at those markets. The distinction between conditional and unconditional distributions of returns is made by considering the technical trading signals derived from technical tools, x_i, presented in Table 8.2. More precisely, conditional distributions are those that include returns generated by adopting an x_i on the dataset considered. The following analysis uses logarithmic, daily returns and is based on the corresponding analysis conducted in Brock et al. (1992). Raw returns are preferable than excess returns, and more appropriate for short term trading rules, since risk premia are likely to have a long periodicity relative to the holding period (Sweeney 1986). In addition we evade the joint hypothesis problem concerning the choice of the model to use for the calculation of excess returns. For example the validity of two of the most known asset pricing models (*Capital Asset Pricing Model*-CAPM and *Arbitrage Pricing Theory*-APT) suffers from strong theoretical and empirical limitations. On the one hand, CAPM is untestable due to the unobservability of the market portfolio (Roll 1977). On the other hand, for the APT it is not necessary to identify the market portfolio, but it is not specified what the underlying factors are (unlike the CAPM, which collapses all the macroeconomic factors into the market portfolio).

Empirical results for the NIKKEI index (idx_{16}), for 5 days holding period are tabulated in Table 8.3. Each technical tool, x_i, signals long and short positions and distributions of daily logarithmic returns generated are compared with the unconditional return distributions. Table 8.3 presents two rows of results for each $\mathrm{x_i}$. Focusing on $\mathrm{x_1}$, the short term SMA signals 326 long and short positions. However long and short positions last for 1,096 and 1,090 days, and produce 3.62×10^{-4} and -8.96×10^{-4} mean daily returns respectively. B-S is the mean daily return from joining the distributions of returns generated from both long and short positions, after changing the sign of the "Sell" distribution. The calculation of the B-S mean return for x_1 is presented in Eq. (8.1).

$$6.28 \times 10^{-4} = \frac{1096 \times 3.62 \times 10^{-4} + 1090 \times 8.96 \times 10^{-4}}{1096 + 1090} \tag{8.1}$$

It should be mentioned that (Brock et al. 1992) use "B-S" in order to express the difference between the mean daily buy and sell returns and then test the null hypothesis of zero equality in order to compare returns resulted from long and

Table 8.3 Ordinary statistical assessment for the NIKKEI index

x_i	N(Buy)	N(Sell)	Buy	Sell	B-S	B > 0	S > 0	B-S > 0
x_1	326	326	3.62×10^{-4}	-8.96×10^{-4}	6.28×10^{-4}	0.5246	0.4890	0.5178
	1,096	1,090	(0.1346)	(**0.0511**)	(**0.0135**)	(**0.0514**)	(0.2336)	(**0.0476**)
x_2	146	146	3.58×10^{-4}	-7.03×10^{-4}	5.29×10^{-4}	0.5160	0.5033	0.5065
	469	461	(0.2147)	(0.2008)	(**0.0893**)	(0.2443)	(0.5556)	(0.3470)
x_3	48	47	-2.41×10^{-5}	-9.47×10^{-4}	4.38×10^{-4}	0.4678	0.5032	0.4816
	171	155	(0.4643)	(0.2663)	(0.2464)	(0.7999)	(0.5320)	(0.7469)
x_4	50	50	-8.59×10^{-5}	-4.97×10^{-4}	2.08×10^{-4}	0.4770	0.4938	0.4917
	239	243	(0.4933)	(0.3491)	(0.3158)	(0.7616)	(0.4237)	(0.6422)
x_5	167	166	5.82×10^{-4}	4.51×10^{-4}	6.74×10^{-5}	0.5432	0.5312	0.5061
	694	689	(**0.0965**)	(0.8173)	(0.3418)	(**0.0114**)	(0.9493)	(0.3238)
x_6	130	135	3.71×10^{-4}	-2.16×10^{-4}	2.92×10^{-4}	0.5168	0.4911	0.5127
	596	621	(0.2557)	(0.4078)	(0.1874)	(0.2063)	(0.3295)	(0.1871)
x_7	181	167	4.84×10^{-4}	8.21×10^{-5}	2.31×10^{-4}	0.5348	0.5329	0.5046
	546	441	(0.2091)	(0.5916)	(0.2685)	(**0.0519**)	(0.9164)	(0.3873)
x_8	358	357	-2.78×10^{-5}	-2.23×10^{-4}	1.00×10^{-4}	0.5085	0.5179	0.4950
	1,119	1,170	(0.4349)	(0.3919)	(0.2680)	(0.2850)	(0.8903)	(0.6846)
x_9	91	83	-2.62×10^{-4}	3.97×10^{-4}	-3.27×10^{-4}	0.5192	0.5332	0.4938
	416	392	(0.5763)	(0.7384)	(0.6533)	(0.2164)	(0.9054)	(0.6375)
x_{10}	192	138	6.38×10^{-5}	-2.37×10^{-4}	1.37×10^{-4}	0.5403	0.5239	0.5130
	794	586	(0.3427)	(0.4328)	(0.2887)	(**0.0116**)	(0.8763)	(0.1663)
x_{11}	158	173	5.64×10^{-4}	-7.40×10^{-4}	6.54×10^{-4}	0.5242	0.5263	0.4984
	599	627	(0.1238)	(0.1751)	(**0.0490**)	(0.1180)	(0.9062)	(0.5455)
x_{12}	8	6	-9.79×10^{-4}	2.41×10^{-3}	-1.59×10^{-3}	0.4500	0.6000	0.4286
	40	30	(0.7113)	(0.8334)	(0.8541)	(0.7365)	(0.8633)	(0.8840)

(continued)

Table 8.3 (continued)

x_i	N(Buy)	N(Sell)	Buy	Sell	B-S	B > 0	S > 0	B-S > 0
x_{13}	0	1		-1.06×10^{-2}	1.06×10^{-2}		0.4000	0.6000
	0	5		(0.1860)	(0.1821)		(0.3274)	(0.3274)
x_{14}	1	0	9.93×10^{-3}		9.93×10^{-3}	0.8000		0.8000
	5	0	(**0.0359**)		(**0.0359**)	(**0.0899**)		(**0.0899**)
x_{15}	3	0	-1.10×10^{-3}		-1.10×10^{-3}	0.4667		0.4667
	15	0	(0.6638)		(0.6638)	(0.6019)		(0.6019)
x_{16}	1	1	2.63×10^{-3}	1.30×10^{-2}	-5.19×10^{-3}	0.4000	0.6000	0.4000
	5	5	(0.2929)	(0.6201)	(0.6017)	(0.6726)	(0.6726)	(0.7365)
x_{17}	0	2		-5.76×10^{-3}	5.76×10^{-3}		0.4000	0.6000
	0	10		(0.1224)	(0.1150)		(0.2635)	(0.2635)
x_{18}	11	15	-1.42×10^{-3}	-4.21×10^{-3}	1.67×10^{-3}	0.5091	0.4925	0.5082
	55	67	(0.7991)	(0.1264)	(0.1990)	(0.4464)	(0.4514)	(0.4282)

x_i technical tool (see Table 8.2), *N(Buy)* and *N(Sell)* number of generated buy and sell signals along with the number of days of long and short positions accordingly, *Buy*, *Sell* and *B-S* mean daily returns from long, short, and combined positions respectively, $B > 0$, $S > 0$ and $B\text{-}S > 0$ fractions of returns greater than zero, numbers is brackets are p-values from one tail tests (t-tests for *Buy*, *Sell* and *B-S* and z-tests for $B > 0$, $S > 0$ and $B\text{-}S > 0$), p-values lower than 0.1 are in bold

short signals. Contrary, "B-S" is used here as the overall mean daily return generated from a particular x_i, including both long and short signals, in order to test the null hypothesis that this return is not greater than the unconditional mean daily return of the initial price series. Buy this the overall performance of each trading rule is assessed, considering buy and sell signals of a particular x_i simultaneously. Below them, the corresponding p-values are reported in brackets, which result by implementing one-tailed paired student's t-test (right tail for Buy and left tail for Sell returns). Results regarding x_1 indicate that buy and sell signals generate positive and negative mean daily returns. However only the latter are statistically significant lower than the unconditional mean for a significance level of 10 % (p-value is 0.0511). Finally, regarding the overall performance of x_1, the p-value of 0.0135 suggests that long and short signals combined produce a mean daily return significantly greater than the unconditional mean return at a 5 % level.

It is also examined the ability of each trading rule to generate "buy", "sell" and "buy-sell" signals which in turn produce positive or negative returns, irrespective of the size of the returns. To do so, for each case, the fractions of trading days where positive returns are generated are calculated, and it is assessed whether they differ significantly from a 50 % pure chance. For this reason each day a position is opened is treated as a Bernoulli trial where two possible outcomes can result; success (the realization of a positive return) and a failure (the realization of a negative return). p-values are calculated to estimate the significance at which the observed fractions are different than a fair 50 % level. Advocates of TA would expect fractions observed at "Buy" and "Buy-Sell" signals to be significantly greater than 50 %, whilst "Sell" fractions to be significantly lower than a 50 % "fair" level. Referring again to Table 8.3, the corresponding fractions of buy, sell and buy-sell signals, for x_1, have values of 0.5246, 0.4890 and 0.5178 respectively. However, only B > 0 and B-S > 0 are statistically significant as indicated by their p-values (0.0514 and 0.0476) for a significance level of 10 % and 5 % respectively. Thus, it could be argued that, without considering statistical significance, the first trading rule behaves overall as an advocate of TA would expect. It produces mean daily returns with the expected sign (positive for Buy and B-S and negative for Sell) and fractions of realized returns with the expected sign greater (lower) than 50 % for Buy and B-S (Sell). By scrutinizing further these results, and considering the statistical significance, it seems that buy signals generate significantly positive returns more frequently than 50 % of time, but the size of this returns is not sufficiently large in order to be significantly greater than the unconditional mean return. Regarding sell signals, although their fraction of producing negative returns is not significantly greater than 50 %, the size of realised negative returns is large enough to characterize mean Sell returns significantly lower than the unconditional mean at a 10 % level. This characteristic can be attributed to a phenomenon in asset returns observed by Black (1976) where, as Brock et al. (1992) quote, negative returns were generally followed by larger volatility than positive returns. This is a first indication that distributional characteristics of asset returns differ from normality, and raise the necessity to implement a bootstrap assessment (see Sect. 8.4).

Furthermore, there are cases where some technical patterns were not identified in the NIKKEI index. In particular, x_{13} and x_{17} did not produce any buy signals, whereas x_{14} and x_{15} were not identified in their "bearish" version. The performance of technical patterns also varies from case to case. For example although x_{11} generates buy, sell and buy-sell mean returns with the expected by TA sign, mean returns generated by x_{12} have the opposite than expected sign! It is also worth mentioning that patterns are observed much rarer than signals generated by technical indicators, a case which also corroborates technicians' arguments. The small sample size of trading returns in such cases may violate the required conditions of many common statistical tests to be adopted. Thus ordinary statistical tests on these cases should be considered as supplementary rather than robust. Bootstrap method applied in Sect. 8.4 may deal also with this problem.

As expected, the performance among various technical rules varies and is not consistent among different sub periods. Since it is generally argued in the literature that TA is used for short-term forecasting (Gehrig and Menkhoff 2006), we will subsequently present overall results by setting the HP value equal to one. Table 8.4 provides an overall picture of the technical rules' performance. In particular, 18 technical rules implemented on 18 price series, give a total of 324 cases. From these cases the number of instances, *N*, that generated positive and negative mean returns was counted. The number of cases not identified is also provided which refer solely on technical patterns. Finally, the number of cases, *N**, that are statistically significant at 10 % level with their corresponding fractions are also tabulated. A technician would expect positive average returns from "bullish" and "combined" signals and negative from "short" signals.

Initial empirical results fulfill the aforesaid expectation, in the sense that trading signals generate more often returns with the expected sign; positive mean returns for buy and buy-sell signals and negative returns after sell signals. However when focusing only on significant cases, the picture exacerbates. For example, regarding

Table 8.4 Summary results for mean returns

	N	*pct.*	*N**	*pct.**
Bullish >0	153	57.95 %	40	15.15 %
Bullish <0	111	42.05 %		
Bullish: Not identified	60			
Bearish >0	118	46.64 %		
Bearish <0	135	53.36 %	36	14.23 %
Bearish: Not identified	71			
Combined >0	145	52.54 %	43	15.58 %
Combined <0	131	47.46 %		
Combined: Not identified	48			

Bullish (Bearish) Versions of trading rules that generate buy (sell) signals, *Combined* Trading rules that generate both buy and sell signals, *>0 (<0)* Generation of positive (negative) mean returns, *N* Number of rules assigned to each category, *pct.* Percentage
*Indicates statistically significant cases for a significance level of 10 %

buy signals, 57.95 % of all valid cases generate positive mean returns but this figure drops to 15.15 % when considering a significance level of 10 %. Similar are the results for sell and buy-sell signals. 53.36 % of bearish trading rules generate negative mean returns but the significant ones represent a ratio of 14.23 %. Finally, the overall picture of the examined trading rules is that 52.54 % of all cases generate positive mean returns with the significance ratio dropping to 15.58 %.

Table 8.5 reports the breakdown of statistically significant cases presented in Table 8.4. For each pair of x_i and idx_i a binary indexation $a:b:c$ is used to spot significant cases, where $a, b, c \in \{1, 0, -\}$. Values of a, b and c correspond bullish, bearish and combined versions for each x_i and signify significant cases with 1 and non-significant with 0. Cases where the corresponding technical pattern was not identified are marked with "–". Finally the sums across each x_i (last column) and idx_i (last row) are tabulated.

Results indicate that x_1 generates statistically significant returns in more financial indices than any other trading rule does. It is also apparent that x_1, which is a short term SMA, outperforms, as expected, x_2 and x_3 which are medium and long term SMA. This is aligned with TA's assertion that short term parameters should be used in technical indicators, when the user's intention is to make predictions in the short run (recall that a 1-day holding period is used).

In addition, bibliography provides many examples where technical patterns are dominated by simpler trading rules like the SMA. For instance (Chang and Osler 1999) compared the profitability and efficiency of the HS pattern to that of SMA and momentum strategies and found that although trading rules based on the HS is profitable, it underperforms compared to simpler technical rules. These findings are also aligned with the results reported in Table 8.5. Finally, the performance of x_5 and x_{10} is also notable, whereas x_7 did not produce significant profitability at any price series examined.

Focusing on technical patterns, x_{11}, x_{12} and x_{18} produce signals (i.e. they are identified) in every price series tested, which gives them more chances of occurrence, with the first two performing better. However, remaining patterns (x_{13}, x_{14}, x_{15}, x_{16} and x_{17}) are not identified on every index, but they do produce at cases significant profits.

Subsequently the assessment focuses on the ability of a technical rule to predict the signs of generated returns without considering their size. For bullish, bearish and combined versions of each technical rule presented in Table 8.2, the number of instances where generated returns are positive with a frequency greater (less) than 50 % of time were counted (Table 8.6). Results indicate that every version, even bearish ones, produces positive returns most of the times. Conditioning each version of technical rule and the price series tested, the realization of each return is treated as a Bernoulli trial in which is a random event with a 50 % probability of success. One tailed z-tests are implemented (right for Fraction ≥ 0.5 and left for Fraction ≤ 0.5) to isolate significant cases for a 10 % significance level. Initial percentages drop, as expected, and the outcome of this process is 25.38 %, 3.95 % and 16.67 % for bullish, bearish and combined versions of each x_i respectively. The first and last figures are greater than the corresponding ones reported in Table 8.4.

Table 8.5 Breakdown of statistically significant cases

	idx_1	idx_2	idx_3	idx_4	idx_5	idx_6	idx_7	idx_8	idx_9	idx_{10}	idx_{11}	idx_{12}	idx_{13}	idx_{14}	idx_{15}	idx_{16}	idx_{17}	idx_{18}	Total
x_1	1:0:0	1:0:0	0:0:0	1:1:1	1:1:1	0:1:0	0:0:0	0:0:0	1:0:1	1:0:0	1:1:1	0:1:1	0:0:0	0:0:0	1:0:1	0:0:0	1:0:1	0:0:0	**9:5:7**
x_2	0:0:0	0:0:0	0:0:0	0:1:1	0:1:1	0:0:0	0:0:0	0:0:0	0:0:0	0:0:0	0:0:0	0:0:0	0:0:0	0:1:1	0:0:0	0:0:0	0:0:0	0:0:0	**0:3:3**
x_3	0:0:0	0:0:0	0:0:0	0:1:1	0:1:1	0:1:1	0:0:0	0:0:0	0:0:0	0:0:0	0:1:0	0:0:0	0:0:0	0:0:0	0:0:0	0:0:0	0:0:0	0:0:0	**0:4:3**
x_4	0:0:0	0:0:0	0:0:0	0:1:0	0:0:0	0:0:0	0:0:0	0:0:0	0:0:0	0:0:0	1:0:0	0:0:0	0:0:0	0:0:0	0:0:0	0:1:0	0:0:0	0:0:0	**1:2:0**
x_5	0:0:0	0:0:0	0:1:0	0:1:1	1:0:1	1:0:0	0:0:0	0:0:0	0:0:0	0:0:0	0:0:0	1:0:1	0:0:1	0:1:0	1:0:1	0:0:0	1:1:1	1:0:0	**6:4:6**
x_6	0:0:0	0:0:0	0:0:0	0:0:0	0:0:0	0:0:0	1:0:1	0:0:0	0:0:0	0:0:0	0:1:0	0:0:0	0:0:0	0:0:0	0:0:0	0:0:0	0:0:0	0:0:0	**1:1:1**
x_7	0:0:0	0:0:0	0:0:0	0:0:0	0:0:0	0:0:0	0:0:0	0:0:0	0:0:0	0:0:0	0:0:0	0:0:0	0:0:0	0:0:0	0:0:0	0:0:0	0:0:0	0:0:0	**0:0:0**
x_8	0:0:0	0:0:0	0:1:0	1:1:1	0:1:1	0:0:0	0:0:0	0:0:0	0:0:0	0:0:0	0:1:0	0:0:0	0:0:0	0:0:0	0:0:0	0:0:0	0:1:0	0:0:0	**1:5:2**
x_9	0:0:0	0:0:0	0:0:0	0:0:0	0:0:0	1:0:1	0:0:0	0:0:0	0:0:0	0:0:0	0:0:0	0:0:0	0:0:0	0:0:0	0:0:0	0:0:0	0:0:0	0:0:0	**1:0:1**
x_{10}	0:0:0	0:0:0	1:0:0	0:0:0	1:1:1	1:0:1	0:0:0	0:0:0	0:0:0	0:0:0	0:0:0	0:0:0	0:0:0	0:0:0	1:1:1	1:1:1	1:0:1	0:0:0	**6:3:5**
x_{11}	0:0:0	0:0:0	0:0:0	0:0:0	0:0:0	0:0:0	0:0:0	1:0:1	0:0:0	0:0:0	1:1:1	0:0:1	0:0:1	0:0:0	0:1:1	1:0:1	0:0:0	0:0:0	**3:2:6**
x_{12}	0:0:0	1:0:1	1:0:0	1:1:1	0:0:0	0:1:1	0:0:0	0:0:0	0:0:0	0:0:0	0:0:0	0:0:0	0:0:0	0:0:0	0:0:0	0:0:0	0:0:0	1:1:1	**4:3:4**
x_{13}	0:–:0	0:–:0	–:0:0	–:–:–	–:–:–	–:–:–	–:–:–	–:0:0	–:–:–	–:–:–	–:–:–	–:0:0	0:–:0	–:0:0	–:0:0	–:1:0	–:–:–	–:0:0	**0:1:0**
x_{14}	–:–:–	0:0:0	0:0:0	–:–:–	1:0:0	–:–:–	0:–:0	–:–:–	0:–:0	0:–:0	0:0:0	0:–:0	0:0:0	–:–:–	–:–:–	0:–:0	–:–:–	0:–:0	**1:0:0**
x_{15}	–:–:–	–:–:–	1:0:0	–:–:–	–:–:–	–:–:–	1:–:1	0:–:0	0:–:0	0:–:0	1:–:1	0:–:0	0:–:0	0:–:0	0:–:0	0:–:0	0:–:0	0:–:0	**3:0:2**
x_{16}	–:–:–	–:–:–	–:–:–	–:–:–	0:–:0	–:0:0	–:0:0	–:–:–	–:–:–	–:–:–	–:–:–	–:–:–	–:–:–	–:–:–	–:–:–	1:1:1	–:–:–	–:–:–	**1:1:1**
x_{17}	–:–:–	–:–:–	–:–:–	–:1:0	–:–:–	–:–:–	–:–:–	–:–:–	–:–:–	–:–:–	–:–:–	–:–:–	–:–:–	–:–:–	–:–:–	–:0:0	1:–:1	–:1:0	**1:2:1**
x_{18}	0:0:0	0:0:0	0:0:0	0:0:0	1:0:1	0:0:0	0:0:0	0:0:0	1:0:0	0:0:0	0:0:0	0:0:0	0:0:0	0:0:0	0:0:0	0:0:0	0:0:0	0:0:0	**2:0:1**
Total	**1:0:0**	**2:0:1**	**3:2:0**	**3:8:6**	**5:5:7**	**3:3:4**	**2:0:2**	**1:0:1**	**2:0:1**	**1:0:0**	**4:5:3**	**1:1:3**	**0:0:2**	**0:2:1**	**3:2:4**	**3:4:3**	**4:2:4**	**2:2:1**	**40:36:43**

x:*y*:*z* binary indexing of significant cases (significance level of 10 %). "–" indicates that the corresponding pattern was not identified

Table 8.6 Summary results for returns' signs

	N	*pct.*	*N**	*pct.**
Panel A: Bullish				
Fraction ≥ 0.5	188	71.21 %	67	25.38 %
Fraction < 0.5	76	28.79 %		
Not identified	60			
Panel B: Bearish				
Fraction > 0.5	152	60.08 %		
Fraction ≤ 0.5	101	39.92 %	10	3.95 %
Not identified	71			
Panel C: Combined				
Fraction ≥ 0.5	156	56.52 %	46	16.67 %
Fraction < 0.5	120	43.48 %		
Not identified	48			

Fraction > 0.5 (Fraction < 0.5) Cases where signals generate positive returns with a frequency greater (less) than 50 % of time, *Bullish (Bearish)* Versions of trading rules that generate buy (sell) signals, *Combined* Trading rules that generate both buy and sell signals, *N* Number of rules assigned to each category, *pct.* Percentage

*Indicates statistically significant cases for a significance level of 10 %

Bullish and the overall performance of trading rules examined predict more frequently the sign of returns compared with the cases where the size of returns is also considered. An interesting finding is that there were more cases of bearish rules generating returns with a mean lower than the unconditional mean, than predicting returns with a negative sign most of the times. This implies that negative returns predicted by these rules were great in absolute values; a finding that can be attributed to the aforementioned black phenomenon.

Table 8.7 presents the ranking of trading rules according to the mean trading returns they have generated after buy, sell and buy-sell signals, considering the whole dataset. It is shown that the in the first ranks we mainly identify cases of technical patterns. However, we should consider that technical indicators generally produce significantly more trading signals than technical patterns do, which "forces" mean conditional trading returns to move towards the unconditional means. Finally, there are cases whereby technical tools produce mean trading returns with different sign than the one expected. More precisely, $x_i, i \in \{3, 4, 7, 8, 9, 15\}$ produce negative mean trading returns after buy signals, $x_i, i \in \{6, 7, 9, 14, 18\}$, produce positive mean returns after sell signals and $x_i, i \in \{3, 4, 7, 9, 14, 15, 18\}$ produce negative mean returns when their long and short trading signals are combined.

The overall performance of all technical tools implemented on all market indices is being illustrated in Fig. 8.1 by using different HP values. Mean daily returns for Buy, Sell and Buy-Sell signals are presented in Fig. 8.1a. For comparison reasons, returns generated by Sell signals are multiplied with −1. Figure 8.1b illustrates the

Table 8.7 Ranking trading rules

Rank	x_i	Buy	x_i	Sell	x_i	Buy-Sell
1	x_{17}	4.39×10^{-2}	x_{16}	-3.13×10^{-2}	x_{16}	2.68×10^{-2}
2	x_{16}	1.31×10^{-2}	x_{17}	-1.06×10^{-2}	x_{17}	2.98×10^{-3}
3	x_{12}	2.17×10^{-3}	x_{15}	-5.09×10^{-3}	x_{12}	1.64×10^{-3}
4	x_1	1.12×10^{-3}	x_{12}	-1.10×10^{-3}	x_{13}	1.35×10^{-3}
5	x_{18}	1.10×10^{-3}	x_{13}	-8.52×10^{-4}	x_1	8.56×10^{-4}
6	x_{13}	1.01×10^{-3}	x_{11}	-6.55×10^{-4}	x_5	6.70×10^{-4}
7	x_5	8.58×10^{-4}	x_1	-5.91×10^{-4}	x_{11}	6.49×10^{-4}
8	x_6	8.40×10^{-4}	x_8	-5.64×10^{-4}	x_{10}	4.99×10^{-4}
9	x_{11}	6.43×10^{-4}	x_{10}	-5.27×10^{-4}	x_8	2.24×10^{-4}
10	x_{14}	6.32×10^{-4}	x_5	-4.82×10^{-4}	x_6	2.03×10^{-4}
11	x_{10}	4.84×10^{-4}	x_3	-1.11×10^{-4}	x_2	1.81×10^{-4}
12	x_2	2.70×10^{-4}	x_4	-9.28×10^{-5}	x_{15}	-6.43×10^{-6}
13	x_8	-1.16×10^{-4}	x_2	-9.25×10^{-5}	x_3	-8.76×10^{-5}
14	x_{15}	-1.71×10^{-4}	x_6	2.19×10^{-4}	x_7	-3.40×10^{-4}
15	x_7	-2.31×10^{-4}	x_7	4.43×10^{-4}	x_4	-5.96×10^{-4}
16	x_3	-2.83×10^{-4}	x_9	4.45×10^{-4}	x_9	-7.39×10^{-4}
17	x_9	-9.81×10^{-4}	x_{18}	1.68×10^{-3}	x_{18}	-1.04×10^{-3}
18	x_4	-1.29×10^{-3}	x_{14}	8.64×10^{-3}	x_{14}	-1.81×10^{-3}

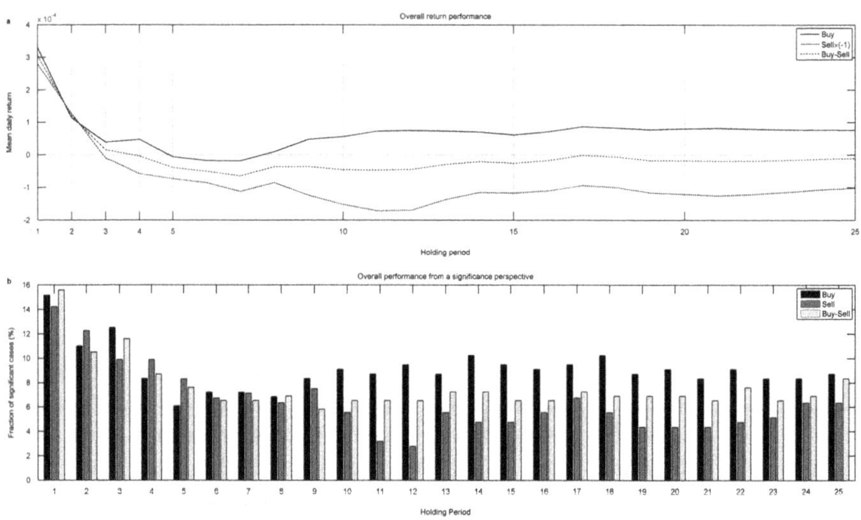

Fig. 8.1 Overall performance for different HP values

fractions of significant cases spotted. For instance, the first group of bars represents the *pct.** values tabulated in Table 8.4. The rest bar-groups depict generated *pct.** values from different HP values. TA seems to perform best when short term holding periods are used. On average, using a 1-day HP generates the greatest returns and significant fractions for bullish, bearish and combined versions of the examined trading rules. However, increasing the holding period, decreases initially TA's overall performance and subsequently it fluctuates less. These findings are aligned with the assortment that TA should be preferred when the prediction period is short run. Regarding the superiority of bullish trading rules, recall that a similar fining has been reported in Chap. 4 where support levels, which generate long signals, outperformed resistance levels, which in turn they are considered bearish rules (Zapranis and Tsinaslanidis 2012).

8.4 A Bootstrap Assessment

Before proceeding to the bootstrap methodology, Table 8.8 presents the distributional characteristics of the dataset considered by using the following statistics: mean, standard deviation, skewness, kurtosis and the Kolmogorov-Smirnoff (*KS*-stat) test for normality. Autocorrelation coefficients $\rho(\ell)$ for lags $\ell = 1, 2, \cdots, 5$ days, as well as Ljung-Box-Pierce statistics $LBQ(\ell)$ for lags $\ell = 6$, 12 and 24 are also tabulated.

Let $r_t = \ln(P_t/P_{t-1}) = \ln P_t - \ln P_{t-1}$ be the logarithmic daily return, with P_t being the index price on time t. The sample bias-corrected skewness G_1 is defined as

$$G_1 = \frac{\sqrt{N(N-1)}}{N-2} g_1, \tag{8.2}$$

where N is the sample size and

$$g_1 = \frac{\frac{1}{N}\sum_{t=1}^{N} (r_t - \bar{r})^3}{\left(\sqrt{\frac{1}{N}\sum_{t=1}^{N} (r_t - \bar{r})^2}\right)^3} \tag{8.3}$$

(Joanes and Gill 1998).

The significance of skewness is subsequently assessed, i.e. the probability that the population is skewed by estimating the test statistic $Z_{G_1} = G_1/SES$ where SES is the standard error of skewness (Eq. 8.4) (Cramer 1997).

$$SES = \sqrt{\frac{6N(N-1)}{(N-2)(N+1)(N+3)}} \tag{8.4}$$

With this statistic a two-tailed test is performed, testing the null hypothesis $H_0 : G_1 = 0$ against the alternative $H_1 : G_1 \neq 0$. Table 8.8 reports that idx_i, $i \in \{2, 4, 5, 6, 8, 13, 16, 18\}$ and idx_i, $i \in \{12, 14\}$ are skewed (i.e. skewness is different than zero) for a significance level of 1 % and 5 % respectively.

Table 8.8 Statistical properties

	N	Mean	St. dev	Skewness	Kurtosis	$\rho(1)$	$\rho(2)$	$\rho(3)$	$\rho(4)$	$\rho(5)$	Bartlett	$LBQ(6)$	$LBQ(12)$	$LBQ(24)$	KS-stat
idx_1	3900	1.24×10^{-4}	0.0121	−0.026	10.272**	−0.056**	−0.050**	0.014	0.005	−0.025	0.016	26.366**	44.357**	67.099**	0.079**
idx_2	3900	9.27×10^{-5}	0.0130	−0.163**	10.460**	−0.063**	−0.048**	0.000	−0.001	−0.029	0.016	28.607**	46.672**	75.125**	0.077**
idx_3	3900	1.59×10^{-4}	0.0174	−0.002	7.534**	−0.018	−0.044*	−0.001	0.002	−0.019	0.016	11.976	21.791*	45.077**	0.073**
idx_4	3900	1.54×10^{-4}	0.0117	−0.781**	10.889**	0.017	−0.051**	0.014	0.013	−0.058**	0.016	29.045**	35.155**	46.538**	0.080**
idx_5	3900	5.46×10^{-4}	0.0149	0.117**	8.449**	0.102**	−0.031	−0.032*	0.012	−0.017	0.016	50.258**	62.562**	77.708**	0.069**
idx_6	3900	4.48×10^{-4}	0.0207	0.541**	18.206**	0.013	−0.018	−0.036*	−0.025	−0.011	0.016	11.697	44.083**	62.319**	0.063**
idx_7	3900	9.83×10^{-7}	0.0155	−0.026	7.053**	−0.009	−0.040*	−0.069**	0.043*	−0.053**	0.016	45.004**	53.524**	74.502**	0.066**
idx_8	3900	3.04×10^{-5}	0.0126	−0.133**	8.565**	−0.025	−0.056**	−0.084**	0.065**	−0.057**	0.016	76.539**	87.907**	114.984**	0.067**
idx_9	3900	4.20×10^{-5}	0.0153	−0.009	7.454**	−0.011	−0.039*	−0.069**	0.032*	−0.053**	0.016	41.258**	51.236**	67.393**	0.065**
idx_{10}	3900	1.41×10^{-4}	0.0159	−0.075	6.918**	0.002	−0.027	−0.036*	0.042*	−0.038*	0.016	22.805**	28.248**	45.836**	0.066**
idx_{11}	3900	1.76×10^{-5}	0.0155	−0.002	7.935**	0.021	−0.041*	−0.060**	0.016	−0.029	0.016	28.016**	33.249**	51.044**	0.062**
idx_{12}	3900	-1.18×10^{-4}	0.0156	−0.086*	7.354**	0.008	−0.011	−0.056**	0.068**	−0.069**	0.016	49.863**	59.981**	86.871**	0.073**
idx_{13}	3900	-5.64×10^{-5}	0.0154	−0.119**	8.523**	0.003	−0.009	−0.074**	0.051**	−0.065**	0.016	50.711**	67.966**	86.169**	0.074**
idx_{14}	3900	1.47×10^{-4}	0.0161	0.088*	6.435**	−0.002	−0.035*	−0.048**	0.005	−0.043*	0.016	27.302**	33.695**	67.065**	0.057**
idx_{15}	3900	1.65×10^{-5}	0.0125	−0.074	8.855**	0.040*	−0.039*	−0.038*	0.044*	−0.074**	0.016	53.303**	62.245**	76.544**	0.074**
idx_{16}	3900	-9.33×10^{-5}	0.0147	−0.580**	8.757**	0.002	−0.022	−0.036*	0.009	−0.005	0.016	8.604	14.127	24.368	0.056**
idx_{17}	3900	2.02×10^{-4}	0.0165	−0.010	9.727**	0.023	−0.014	0.011	−0.021	−0.039*	0.016	11.805	18.770	37.255*	0.074**
idx_{18}	3900	1.51×10^{-4}	0.0101	−0.489**	8.956**	−0.026	−0.005	−0.022	−0.008	0.001	0.016	6.198	11.168	18.984	0.068**

N Number of observations, *St. dev* standard deviation, $\rho(\ell)$ autocorrelation coefficient for lag ℓ, *Bartlett* Bartlett standard $= 1/\sqrt{N}$, $LBQ(\ell)$ Ljung-Box-Pierce Q statistic for lag ℓ, *KS-stat* Kolmogorov-Smirnov statistic
*Indicates statistical significance at the 5 % level
**Indicates statistical significance at the 1 % level

Unconditional distributions of the rest indices were not found to be skewed. However this might be attributed to the fact that samples considered are relative large which constitutes their distribution skewness to tend towards zero.

Sample bias-corrected Kurtosis G_2 is calculated as

$$G_2 = \frac{N-1}{(N-2)(N-3)}((N+1)g_2 - 3(N-1)) + 3, \tag{8.5}$$

where

$$g_2 = \frac{\frac{1}{N}\sum_{t=1}^{N}(r_t - \bar{r})^4}{\left(\frac{1}{N}\sum_{t=1}^{N}(r_t - \bar{r})^2\right)^2}. \tag{8.6}$$

The corresponding test statistic is $Z_{G_2} = G_2/SEK$, where SEK is the standard error of kurtosis (Eq. 8.7).

$$SEK = 2SES\sqrt{\frac{N^2-1}{(N-3)(N+5)}} \tag{8.7}$$

Results presented in Table 8.8 show that Kurtosis is statistically significant (at a level of 1 %) different than three in the return distributions of all indices. These values indicate that these distributions are leptokurtic i.e. they exhibit fat tails and high concentration of daily returns around the mean. Thus the KS-test rejects the null of normality in all cases.

Finally, for the majority of the examined price series the "portmanteau" LBQ-test rejects the null hypothesis that returns exhibit no autocorrelation for a fixed number of lags $\ell = 6,\ 12$ and 24. However the null is not rejected for idx_{16} and idx_{18}.

The bootstrap methodology conducted is based on the one presented in Brock et al. (1992) to address the problems that arise due to the non-normal distributional characteristics along with the non-stationary and autocorrelation behaviour the examined price series exhibit. The aim of this analysis is to fit a null model on each original price series. Subsequently, simulated prices series are being produced based on the coefficients estimated, by sampling with replacement their residuals. In order to do that, increments should be independent and identically distributed, i.e. they should not exhibit autocorrelation or heteroskedasticity. Technical trading rules are then applied on both the original series and the simulated ones, and generated results are being compared. In that manner, simulated p-values are extracted which measure the relative predictability of the examined trading rules between the actual and simulated price series. Thus, it can be determined whether trading profits result due to a specific null model.

The simulated price series are based on the three following widely used null models; *autoregressive process of order one* AR(1) (Eq. 8.8), *generalised autoregressive conditional heteroskedasticity in the mean* model GARCH-m (Eqs. 8.9–8.11) and *exponential GARCH* model E-GARCH (Eqs. 8.12–8.14).

The AR(1) model is described in Eq. (8.8).

$$r_t = b + \rho r_{t-1} + \varepsilon_t, \quad |\rho| < 1 \tag{8.8}$$

Here, the return at time t (r_t) is described by a constant b, a proportion ρ of its one-period lagged value and a random shock or disturbance ε_t. In addition, $\{\varepsilon_t\}$ is assumed to be a white noise series with a mean zero and a variance σ_ε^2. Coefficients and residuals result from fitting this model on each actual price series by using ordinary least squares (OLS) method.

The second model used for simulation is the GARCH-m model which is defined as,

$$r_t = \alpha + \gamma h_t + b\varepsilon_{t-1} + \varepsilon_t, \tag{8.9}$$

$$h_t = a_0 + a_1\varepsilon_{t-1}^2 + \beta h_{t-1}, \tag{8.10}$$

$$\varepsilon_t = h_t^{1/2} z_t, \quad z_t \sim N(0,1). \tag{8.11}$$

The conditional variance at time, h_t, is a function of the squared last period's error, ε_{t-1}^2, and the last period's conditional variance, h_{t-1}. The error term ε_t is conditionally normally distributed and serially uncorrelated. In this model the relationship between the return and its volatility is considered. In addition, this model captures the phenomenon of heteroskedasticity, since it implies that periods of high (low) volatility are likely to be followed by periods of high (low) volatility and thus the presence of volatility clusters. Parameters and residuals of this model are estimated using maximum likelihood (Brock et al. 1992).

The last model implemented in this section is the E-GARCH model (Eqs. 8.12–8.14).

$$r_t = a + \gamma r_{t-1} + \beta\varepsilon_{t-1} + \varepsilon_t, \tag{8.12}$$

$$ln\sigma_t^2 = \kappa + G\, log\sigma_{t-1}^2 + \mathrm{A}\left[\frac{|\varepsilon_{t-1}|}{\sigma_{t-1}} - \sqrt{2/\pi}\right] + L\left(\frac{\varepsilon_{t-1}}{\sigma_{t-1}}\right), \tag{8.13}$$

$$\varepsilon_t = \sigma_t z_t, \quad z_t \sim N(0,1). \tag{8.14}$$

This model allows asymmetric effects between positive and negative returns.[1] As Brock, et al. (1992) quote, this model captures the phenomenon in asset returns

[1] The E-GARCH model described in the corresponding equations is not the same with the one developed and presented by Nelson (1991). For more details and descriptions for this version of the E-GARCH model the reader can see (Econometrics Toolbox™ User's Guide R2011b).

observed by Black (1976) where negative returns were generally followed by larger volatility than positive returns. In addition, in this model the natural logarithm of the conditional variance follows an autoregressive process. It uses logged conditional variance to relax the positiveness constraint of model coefficients (Ruey 2002). Parameters of this model are estimated also using maximum likelihood criterion.

These three models are applied on each one of the original price series and the corresponding parameters are estimated. Residuals obtained from AR(1) and standardized residuals produced by GARCH-m and E-GARCH models are resampled with replacement. Estimated coefficients and scrambled residuals are combined and 100 artificial price series are created for each original series.[2] The simulated series keep the same characteristics with the original ones but they are random. Before proceeding to the scrambling of the residuals (or the standardized residuals), two prerequisites must be fulfilled. First, the model must fit well the original series. This can be examined by the statistical significance of the estimated coefficients. The second condition is that increments must be independent and identically distributed (IID). Independency implies no autocorrelation and identically distribution requires homoskedasticity. This criterion is examined by implementing a *Ljung-Box Q-test* (LBQ-test) and an *Engle's ARCH-test* to the residuals obtained from AR(1) and standardized residuals taken from GARCH-m and E-GARCH. The first test evaluates the independence of the increments and the second one assesses if there is any ARCH effect (heteroskedasticity) in them. Only cases where residuals are IID will be considered subsequently into the bootstrap analysis.

Figure 8.2 illustrates the results of *LBQ-tests* and *Engle's ARCH-tests* which are implemented on the generated residuals for 20 lags and at a significance level of 10 %. Regarding the LBQ-test an H value of zero indicates acceptance of the null hypothesis where no autocorrelation exists at the residuals series. Alternatively when H takes the value of one, this implies a rejection of the null hypothesis. Concerning Engle's ARCH-tests, the null hypothesis is that no ARCH effects exist (i.e. there is homoskedasticity) against the alternative of heteroskedasticity existence.

Results of this "filtering" process are shown in Fig. 8.2. For AR(1) model there are only two cases (idx_{16} and idx_{18}) where residuals are independent, but all cases realise heteroskedasticity in their residuals. Thus AR(1) is not qualified to the subsequent bootstrapping methodology. GARCH-m model filters 4 cases where standardized residuals are IID (idx_4, idx_5, idx_{14} and idx_{16}). Finally, from E-GARCH model, five indices satisfy the aforementioned requirements: those filtered by the GARCH-m model plus idx_{18}. Without this "filtering" 54 cases (18 price series times 3 null models) would be qualified to the bootstrap methodology. For each one of these 54 cases 100 artificial series would have to be simulated (5,400 artificial

[2] In Brock et al. (1992) the number of replications used was 500. In this section since 18 indices are examined, due to limitations imposed by computational power the number of replications is reduced to 100.

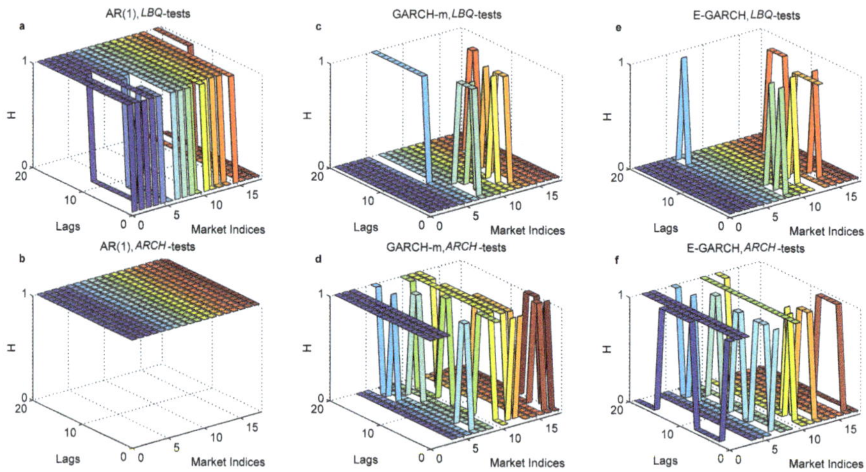

Fig. 8.2 LBQ-tests and Engle's ARCH-tests on residuals obtained from three null models

Table 8.9 GARCH-m coefficients

idx_i	α	b	γ	α_0	β	α_1
idx_4	4.7824×10^{-4}	0.0530	0.5040	1.0583×10^{-6}	0.9136	0.0785
	(2.33*)	(3.08**)	(0.24)	(5.52**)	(173.79**)	(14.53**)
idx_5	5.2963×10^{-4}	0.0945	3.0939	1.9971×10^{-6}	0.9086	0.0840
	(1.84)	(5.28*)	(1.86)	(6.13**)	(159.66**)	(14.34**)
idx_{14}	6.1758×10^{-4}	−0.0108	0.7602	2.1011×10^{-6}	0.9061	0.0874
	(2.22*)	(−0.64)	(0.54)	(5.14**)	(130.53**)	(12.96**)
idx_{16}	1.2771×10^{-4}	0.0217	1.4665	4.7279×10^{-6}	0.8769	0.1020
	(0.37)	(1.2)	(0.76)	(5.96**)	(94.51**)	(13.96**)

Numbers in brackets are the t-statistics
*Indicates statistical significance at 5 % level
**Indicates statistical significance at 1 %

series). Most of the results obtained from such an analysis would not be robust since most cases do not satisfy the aforementioned criteria for the bootstrap analysis. On the contrary, only 9 cases (4 for GARCH-m and 5 for E-GARCH) satisfy these criteria and thus can be used to generate artificial series for evaluation. The overall benefits of this procedure are that these results will be statistically reliable and less computational power will be required for the whole analysis.

Coefficients estimated when the GARCH-m model is fitted are presented in Table 8.9. Coefficients estimated in the variance equation are all significant at the 1 % level. In particular, β and a_1 are highly statistical significant which imply that conditional variance of returns is time varying and autocorrelated. The MA(1) term (coefficient b) is significant only in idx_4 and idx_5 at the 1 % and 5 % level respectively, which signifies the presence of short horizon autocorrelations. Finally,

Table 8.10 E-GARCH coefficients

idx_i	α	γ	β	k	G	A	L
idx_4	2.1728×10^{-4}	−0.0336	0.0913	−0.1323	0.9855	0.1262	−0.0776
	(1.37)	(−0.12)	(0.33)	(−7.36**)	(505.46**)	(13.18**)	(−10.76**)
idx_5	1.7179×10^{-1}	−0.0496	0.1472	−0.1532	0.9819	0.1508	−0.0976
	(2.18*)	(−0.29)	(0.87)	(−8.66**)	(485.22**)	(12.38**)	(−13.53**)
idx_{14}	2.8982×10^{-2}	0.7351	−0.7594	−0.1550	0.9817	0.1592	−0.0853
	(1.24)	(3.59**)	(−3.89**)	(−7.45**)	(409.15**)	(13.82**)	(−12.52**)
idx_{16}	2.1511×10^{-1}	−0.7564	0.7738	−0.2949	0.9656	0.1844	−0.0911
	(−0.33)	(−3.11**)	(3.31**)	(−8.54**)	(245.17**)	(14.1**)	(−11.99**)
Idx_{18}	7.4043×10^{-1}	−0.1050	0.0949	−0.1857	0.9805	0.1337	−0.1098
	(0.66)	(−0.07)	(0.06)	(−8.35**)	(423.34**)	(11.71**)	(−14.8**)

Numbers in brackets are the *t*-statistics
*Indicates statistical significance at 5 % level
**Indicates statistical significance at 1 %

there are no indications that conditional variance and conditional mean are related, with an exception to idx_5 where γ is significant but at the 10 % level.

Table 8.10 presents coefficients estimated by the E-GARCH model. Again, coefficients in the variance equation are all significant at the 1 % level. However it is noteworthy that the leverage coefficient, L, is negative and highly significant which implies that unexpected downward shocks increase the variance. In other words, the existence of the leverage effect, implies that volatility tend to rise (drop) in response to lower (higher) than expected returns.

Table 8.11 presents the results obtained from the bootstrap methodology, when the GARCH-m null model was used, for 1-day holding period. For a particular trading rule x_i let $\mu_b < \widetilde{\mu}_b$ be the fraction of buy mean returns generated from simulated series, $\widetilde{\mu}_b$, larger than the mean buy return achieved in the original series, μ_b. As in Brock et al. (1992), these fractions are treated as simulated "*p*-values" for the evaluation of each trading rule. For instance, a trading rule is said to have statistically significant forecasting power at the 10 % level if more profit is produced on the random series as compared to the original series less than 10 % of the time. Similarly, for buy-sell signals the corresponding fractions reported in Table 8.13 are $\mu_{bs} < \widetilde{\mu}_{bs}$. However, in case of sell signals (Table 8.12) the inequality symbol was reversed $(\mu_s > \widetilde{\mu}_s)$ for consistency purposes, i.e. spotting again significant cases when fractions are lower than a benchmark value. Fractions for the standard deviations of simulated and actual returns are also estimated to examine if any realised profitability is a result of taking higher risk.

Tables 8.11, 8.12, and 8.13 present the results across different indices for buy, sell and combined signals respectively, when the GARCH-m model was used as a null model. Tables 8.14, 8.15, and 8.16 present the corresponding results when the E-GARCH model was used as a null model. These results vary, and are not consistent across different indices, trading rules, and types of trading signals. The majority of trading rules fail to perform significantly different in original series compared to simulated ones. However there are cases where simulated *p*-values are

Table 8.11 Simulation results from GARCH-m bootstrapped replicas; buy signals

	idx_4		idx_5		idx_{14}		idx_{16}	
	$\mu_b < \widetilde{\mu}_b$	$\sigma_b > \widetilde{\sigma}_b$	$\mu_b < \widetilde{\mu}_b$	$\sigma_b > \widetilde{\sigma}_b$	$\mu_b < \widetilde{\mu}_b$	$\sigma_b > \widetilde{\sigma}_b$	$\mu_b < \widetilde{\mu}_b$	$\sigma_b > \widetilde{\sigma}_b$
x_1	0.220	0.260	0.180	0.130	**0.050**	0.160	0.900	0.530
x_2	0.560	0.270	0.380	0.330	0.220	0.300	0.540	0.340
x_3	0.520	0.000	0.580	0.240	0.600	0.460	0.990	0.010
x_4	0.770	0.040	0.930	0.280	0.920	0.410	0.800	0.030
x_5	0.400	0.500	**0.090**	0.310	0.590	0.470	0.810	0.330
x_6	0.300	0.740	0.880	0.660	0.170	0.950	0.160	0.600
x_7	0.850	0.660	0.440	0.780	0.630	0.520	0.300	0.500
x_8	**0.010**	0.420	0.700	0.550	0.970	0.280	0.800	0.630
x_9	0.500	0.940	0.940	0.470	0.860	0.580	0.400	0.150
x_{10}	0.750	0.020	0.510	0.020	0.360	0.030	0.180	0.000
x_{11}	0.470	0.760	0.750	0.160	0.310	0.520	**0.070**	0.080
x_{12}	0.130	0.130	0.340	0.660	0.505	0.263	0.730	0.500
x_{13}								
x_{14}			**0.082**	0.796			0.490	0.510
x_{15}					0.810	0.560	0.338	0.779
x_{16}			0.300	0.900			**0.000**	1.000
x_{17}								
x_{18}	0.840	0.000	0.140	0.090	0.360	0.000	0.670	0.010

Fractions $\mu_b < \widetilde{\mu}_b$ lower than 0.1 are in bold

Table 8.12 Simulation results from GARCH-m bootstrapped replicas; sell signals

	idx_4		idx_5		idx_{14}		idx_{16}	
	$\mu_s > \widetilde{\mu}_s$	$\sigma_s > \widetilde{\sigma}_s$	$\mu_s > \widetilde{\mu}_s$	$\sigma_s > \widetilde{\sigma}_s$	$\mu_s > \widetilde{\mu}_s$	$\sigma_s > \widetilde{\sigma}_s$	$\mu_s > \widetilde{\mu}_s$	$\sigma_s > \widetilde{\sigma}_s$
x_1	**0.090**	0.490	**0.010**	0.330	0.380	0.610	0.410	0.430
x_2	0.130	0.460	**0.070**	0.470	**0.010**	0.160	0.830	0.090
x_3	0.220	0.120	**0.050**	0.850	0.410	0.400	0.740	0.920
x_4	**0.050**	0.460	0.320	0.490	0.940	0.470	**0.000**	0.930
x_5	0.190	0.430	0.710	0.490	**0.090**	0.410	0.450	0.730
x_6	0.380	0.350	0.520	0.240	0.640	0.120	0.800	0.320
x_7	0.220	0.780	0.450	0.560	**0.070**	0.930	0.920	0.970
x_8	0.110	0.410	**0.070**	0.150	0.430	0.450	0.250	0.470
x_9	0.170	0.610	0.580	0.240	0.320	0.400	0.960	0.360
x_{10}	0.430	0.940	**0.080**	0.860	0.930	0.850	**0.020**	0.900
x_{11}	0.450	0.680	0.210	0.700	0.630	0.480	0.320	0.540
x_{12}	0.200	0.050	0.240	0.770	0.860	0.860	0.690	0.250
x_{13}					0.517	0.966	**0.069**	0.828
x_{14}			0.816	0.816				
x_{15}								
x_{16}							**0.000**	1.000
x_{17}	**0.077**	0.923					0.529	0.824
x_{18}	0.750	0.730	0.290	0.800	0.940	0.670	**0.040**	1.000

Fractions $\mu_s > \widetilde{\mu}_s$ lower than 0.1 are in bold

Table 8.13 Simulation results from GARCH-m bootstrapped replicas; buy-sell signals

	idx_4		idx_5		idx_{14}		idx_{16}	
	$\mu_{bs} < \widetilde{\mu}_{bs}$	$\sigma_{bs} > \widetilde{\sigma}_{bs}$	$\mu_{bs} < \widetilde{\mu}_{bs}$	$\sigma_{bs} > \widetilde{\sigma}_{bs}$	$\mu_{bs} < \widetilde{\mu}_{bs}$	$\sigma_{bs} > \widetilde{\sigma}_{bs}$	$\mu_{bs} < \widetilde{\mu}_{bs}$	$\sigma_{bs} > \widetilde{\sigma}_{bs}$
x_1	**0.090**	0.350	**0.040**	0.240	0.120	0.440	0.780	0.440
x_2	0.200	0.340	0.140	0.440	**0.010**	0.180	0.810	0.160
x_3	0.310	0.010	0.160	0.620	0.510	0.400	0.950	0.620
x_4	0.250	0.210	0.770	0.320	0.990	0.390	0.130	0.750
x_5	0.240	0.470	0.240	0.380	0.220	0.440	0.780	0.560
x_6	0.230	0.500	0.760	0.440	0.380	0.730	0.470	0.500
x_7	0.560	0.720	0.430	0.660	0.180	0.820	0.650	0.860
x_8	**0.030**	0.420	0.260	0.300	0.870	0.370	0.580	0.560
x_9	0.270	0.810	0.890	0.340	0.690	0.500	0.800	0.220
x_{10}	0.620	0.550	0.260	0.350	0.780	0.500	**0.020**	0.290
x_{11}	0.450	0.720	0.540	0.400	0.450	0.490	**0.050**	0.260
x_{12}	0.110	0.000	0.260	0.520	0.780	0.660	0.730	0.200
x_{13}					0.622	0.867	**0.059**	0.824
x_{14}			0.348	0.985			0.500	0.364
x_{15}					0.833	0.536	0.329	0.686
x_{16}			0.269	0.885			**0.000**	1.000
x_{17}	**0.048**	0.857					0.455	0.818
x_{18}	0.840	0.780	0.110	0.710	0.970	0.540	**0.090**	0.990

Fractions $\mu_{bs} < \widetilde{\mu}_{bs}$ lower than 0.1 are in bold

significant. For example in Table 8.11 the MACD (x_5) produces mean buy returns in the idx_5 which are greater than those obtained from the simulated series 91 % of time (the p-value is 0.09) and these returns are less volatile 69 % of time (the standard deviation's p-value is 0.31). Similarly, short term MA (x_1) produces significant mean sell returns at the idx_4 and idx_5 with p-values 0.09 and 0.01 respectively. However there are cases where significant p-values in the mean returns may result by the fact that more risk was taken. Such a case can be spotted at Table 8.12 where bearish signals of x_7 perform significantly better at idx_{14} than at simulated series but the $\sigma_s > \widetilde{\sigma}_s$ fraction is 0.93 which implies the presence of higher risk.

Table 8.17 is constructed in the same manner as Table 8.5 in order to identify significant results easier. Panel A is a part of Table 8.5 reporting the results from the ordinary statistically assessment for only the five indices used in the bootstrap methodology. Panels B and C tabulate significant results for GARCH-m and E-GARCH simulations respectively. With one cases where the mean return fraction was significant at the 10 % level (p-value lower than 0.1) and the variance fraction was not greater than 0.9 simultaneously are signified. It is obvious that the majority of significant cases reported in simulated results are also reported by the ordinary statistical assessment. However, the bootstrap methodology reduces the number of statistically significant cases. This suggests that some of the distributional assumptions made at the standard test may have an impact on statistical inferences.

Table 8.14 Simulation results from E-GARCH bootstrapped replicas; buy signals

	idx_4		idx_5		idx_{14}		idx_{16}		idx_{18}	
	$\mu_b < \widetilde{\mu}_b$	$\sigma_b > \widetilde{\sigma}_b$	$\mu_b < \widetilde{\mu}_b$	$\sigma_b > \widetilde{\sigma}_b$	$\mu_b < \widetilde{\mu}_b$	$\sigma_b > \widetilde{\sigma}_b$	$\mu_b < \widetilde{\mu}_b$	$\sigma_b > \widetilde{\sigma}_b$	$\mu_b < \widetilde{\mu}_b$	$\sigma_b > \widetilde{\sigma}_b$
x_1	0.180	0.350	0.150	0.060	**0.060**	0.120	0.900	0.630	0.170	0.500
x_2	0.580	0.320	0.450	0.370	0.180	0.270	0.560	0.460	0.520	0.580
x_3	0.620	0.000	0.580	0.140	0.650	0.490	0.980	0.000	0.170	0.010
x_4	0.860	0.000	0.970	0.310	0.920	0.660	0.870	0.150	0.330	0.220
x_5	0.310	0.670	**0.060**	0.230	0.540	0.610	0.690	0.480	**0.000**	0.770
x_6	0.240	0.900	0.810	0.370	0.180	0.960	0.240	0.420	0.280	0.920
x_7	0.930	0.930	0.450	0.970	0.620	0.610	0.340	0.600	0.990	0.970
x_8	**0.020**	0.660	0.580	0.580	0.980	0.240	0.830	0.850	0.890	0.830
x_9	0.670	0.980	0.920	0.400	0.890	0.600	0.510	0.050	0.890	0.610
x_{10}	0.710	0.260	0.310	0.430	0.150	0.300	**0.050**	0.000	0.510	0.080
x_{11}	0.390	0.910	0.800	0.040	0.300	0.490	**0.060**	0.090	0.920	0.220
x_{12}	**0.060**	0.190	0.290	0.710	0.410	0.230	0.657	0.657	0.260	0.030
x_{13}										
x_{14}			**0.055**	0.600			0.462	0.615	0.273	0.909
x_{15}					0.919	0.570	0.306	0.611	0.583	0.369
x_{16}			0.333	0.815			**0.050**	0.700		
x_{17}										
x_{18}	0.860	0.010	0.120	0.280	0.390	0.030	0.660	0.050	0.540	0.750

Fractions $\mu_b < \widetilde{\mu}_b$ lower than 0.1 are in bold

Table 8.15 Simulation results from E-GARCH bootstrapped replicas; sell signals

	idx_4		idx_5		idx_{14}		idx_{16}		idx_{18}	
	$\mu_s > \widetilde{\mu}_s$	$\sigma_s > \widetilde{\sigma}_s$	$\mu_s > \widetilde{\mu}_s$	$\sigma_s > \widetilde{\sigma}_s$	$\mu_s > \widetilde{\mu}_s$	$\sigma_s > \widetilde{\sigma}_s$	$\mu_s > \widetilde{\mu}_s$	$\sigma_s > \widetilde{\sigma}_s$	$\mu_s > \widetilde{\mu}_s$	$\sigma_s > \widetilde{\sigma}_s$
x_1	**0.080**	0.710	**0.060**	0.310	0.310	0.800	0.450	0.530	0.180	0.880
x_2	**0.090**	0.600	**0.050**	0.400	**0.020**	0.090	0.820	0.050	0.540	0.510
x_3	0.190	0.120	0.110	0.840	0.330	0.400	0.830	0.960	0.880	0.200
x_4	**0.010**	0.450	0.360	0.360	0.910	0.510	**0.000**	1.000	0.380	0.570
x_5	0.120	0.650	0.720	0.520	**0.020**	0.330	0.470	0.840	0.710	0.860
x_6	0.340	0.750	0.560	0.730	0.810	0.460	0.910	0.850	0.760	0.710
x_7	0.200	0.960	0.490	0.570	**0.040**	0.980	0.850	0.990	0.760	0.960
x_8	**0.060**	0.590	0.150	0.120	0.380	0.440	0.260	0.550	0.550	0.550
x_9	0.110	0.960	0.690	0.590	0.470	0.730	0.950	0.810	0.570	0.830
x_{10}	0.410	1.000	0.180	0.630	0.850	0.700	**0.050**	0.670	0.400	0.300
x_{11}	0.400	0.750	0.340	0.670	0.610	0.520	0.240	0.600	0.650	0.130
x_{12}	0.140	0.040	0.182	0.747	0.870	0.760	0.590	0.180	0.230	0.500
x_{13}					0.464	0.750	0.160	0.840	0.651	0.767
x_{14}			0.862	0.966						
x_{15}										
x_{16}							**0.000**	0.929		
x_{17}	**0.000**	1.000					0.389	1.000	**0.067**	0.867
x_{18}	0.770	0.580	0.300	0.800	0.920	0.440	**0.010**	1.000	**0.030**	0.970

Fractions $\mu_s > \widetilde{\mu}_s$ lower than 0.1 are in bold

Table 8.16 Simulation results from E-GARCH bootstrapped replicas; buy-sell signals

	idx_4		idx_5		idx_{14}		idx_{16}		idx_{18}	
	$\mu_{bs} < \widetilde{\mu}_{bs}$	$\sigma_{bs} > \widetilde{\sigma}_{bs}$	$\mu_{bs} < \widetilde{\mu}_{bs}$	$\sigma_{bs} > \widetilde{\sigma}_{bs}$	$\mu_{bs} < \widetilde{\mu}_{bs}$	$\sigma_{bs} > \widetilde{\sigma}_{bs}$	$\mu_{bs} < \widetilde{\mu}_{bs}$	$\sigma_{bs} > \widetilde{\sigma}_{bs}$	$\mu_{bs} < \widetilde{\mu}_{bs}$	$\sigma_{bs} > \widetilde{\sigma}_{bs}$
x_1	**0.050**	0.520	**0.030**	0.170	**0.090**	0.440	0.790	0.570	**0.060**	0.780
x_2	0.280	0.430	**0.090**	0.340	**0.020**	0.110	0.780	0.160	0.560	0.560
x_3	0.290	0.000	0.230	0.620	0.490	0.400	0.960	0.740	0.560	0.070
x_4	0.260	0.150	0.820	0.280	0.970	0.670	0.100	0.920	0.350	0.360
x_5	**0.080**	0.690	0.300	0.390	0.140	0.430	0.680	0.700	**0.090**	0.880
x_6	0.290	0.850	0.800	0.430	0.410	0.890	0.550	0.670	0.520	0.900
x_7	0.570	0.950	0.450	0.900	**0.060**	0.970	0.690	0.970	0.980	0.970
x_8	**0.010**	0.620	0.190	0.340	0.920	0.290	0.540	0.720	0.890	0.740
x_9	0.300	1.000	0.910	0.470	0.920	0.680	0.870	0.230	0.820	0.680
x_{10}	0.570	0.930	0.170	0.470	0.450	0.530	**0.000**	0.350	0.440	0.210
x_{11}	0.370	0.900	0.570	0.180	0.440	0.500	**0.050**	0.330	0.790	0.130
x_{12}	**0.080**	0.020	0.180	0.500	0.710	0.500	0.680	0.230	0.150	0.220
x_{13}					0.528	0.722	0.111	0.639	0.717	0.700
x_{14}			0.515	0.971			0.446	0.538	0.259	0.926
x_{15}					0.897	0.529	0.312	0.610	0.581	0.326
x_{16}			0.375	0.750			**0.000**	1.000		
x_{17}	**0.043**	0.870					0.318	1.000	**0.037**	0.926
x_{18}	0.870	0.750	0.190	0.650	0.940	0.490	**0.050**	0.990	0.100	0.940

Fractions $\mu_{bs} < \widetilde{\mu}_{bs}$ lower than 0.1 are in bold

Table 8.17 Breakdown of statistically significant cases

	Panel A: Traditional tests					Panel B: GARCH-m				Panel C: E-GARCH				
	idx_4	idx_5	idx_{14}	idx_{16}	idx_{18}	idx_4	idx_5	idx_{14}	idx_{16}	idx_4	idx_5	idx_{14}	idx_{16}	idx_{18}
x_1	1:1:1	1:1:1	0:0:0	0:0:0	0:0:0	0:1:1	0:1:1	1:0:0	0:0:0	0:1:1	0:1:1	1:0:1	0:0:0	0:0:1
x_2	0:1:1	0:1:1	0:1:1	0:0:0	0:0:0	0:0:0	0:1:0	0:1:1	0:0:0	0:1:0	0:1:1	0:1:1	0:0:0	0:0:0
x_3	0:1:1	0:1:1	0:0:0	0:0:0	0:0:0	0:0:0	0:1:0	0:0:0	0:0:0	0:0:0	0:0:0	0:0:0	0:0:0	0:0:0
x_4	0:1:0	0:0:0	0:0:0	0:1:0	0:0:0	0:1:0	0:0:0	0:0:0	0:0:0	0:1:0	0:0:0	0:0:0	0:0:0	0:0:0
x_5	0:1:1	1:0:1	0:1:0	0:0:0	1:0:0	0:0:0	1:0:0	0:1:0	0:0:0	0:0:1	1:0:0	0:1:0	0:0:0	1:0:1
x_6	0:0:0	0:0:0	0:0:0	0:0:0	0:0:0	0:0:0	0:0:0	0:0:0	0:0:0	0:0:0	0:0:0	0:0:0	0:0:0	0:0:0
x_7	0:0:0	0:0:0	0:0:0	0:0:0	0:0:0	0:0:0	0:0:0	0:0:0	0:0:0	0:0:0	0:0:0	0:0:0	0:0:0	0:0:0
x_8	1:1:1	0:1:1	0:0:0	0:0:0	0:0:0	1:0:1	0:1:0	0:0:0	0:0:0	1:1:1	0:0:0	0:0:0	0:0:0	0:0:0
x_9	0:0:0	0:0:0	0:0:0	0:0:0	0:0:0	0:0:0	0:0:0	0:0:0	0:0:0	0:0:0	0:0:0	0:0:0	0:0:0	0:0:0
x_{10}	0:0:0	1:1:1	0:0:0	1:1:1	0:0:0	0:0:0	0:1:0	0:0:0	0:0:1	0:0:0	0:0:0	0:0:0	1:1:1	0:0:0
x_{11}	0:0:0	0:0:0	0:0:0	1:0:1	0:0:0	0:0:0	0:0:0	0:0:0	1:0:1	0:0:0	0:0:0	0:0:0	1:0:1	0:0:0
x_{12}	1:1:1	0:0:0	0:0:0	0:0:0	1:1:1	0:0:0	0:0:0	0:0:0	0:0:0	1:0:1	0:0:0	0:0:0	0:0:0	0:0:0
x_{13}	–:–:–	–:–:–	–:0:0	–:1:0	–:0:0	–:–:–	–:–:–	–:0:0	–:1:1	–:–:–	–:–:–	–:0:0	–:0:0	–:0:0
x_{14}	–:–:–	1:0:0	–:–:–	0:–:0	0:–:0	–:–:–	1:0:0	–:–:–	0:–:0	–:–:–	1:0:0	–:–:–	0:–:0	0:–:0
x_{15}	–:–:–	–:–:–	0:–:0	0:–:0	0:–:0	–:–:–	–:–:–	0:–:0	0:–:0	–:–:–	–:–:–	0:–:0	0:–:0	0:–:0
x_{16}	–:–:–	0:–:0	–:–:–	1:1:1	–:–:–	–:–:–	0:–:0	–:–:–	0:0:0	–:–:–	0:–:0	–:–:–	1:0:0	–:–:–
x_{17}	–:1:0	–:–:–	–:–:–	–:0:0	–:1:0	–:0:1	–:–:–	–:–:–	–:0:0	–:0:1	–:–:–	–:–:–	–:0:0	–:1:0
x_{18}	0:0:0	1:0:1	0:0:0	0:0:0	0:0:0	0:0:0	0:0:0	0:0:0	0:0:0	0:0:0	0:0:0	0:0:0	0:0:0	0:0:0
Total	**3:8:6**	**5:5:7**	**0:2:1**	**3:4:3**	**2:2:1**	**1:2:3**	**2:5:1**	**1:2:1**	**1:1:3**	**2:4:5**	**2:2:2**	**1:2:2**	**3:1:2**	**1:1:2**

Table 8.18 GARCH-m aggregate bootstrap results

idx_i	Buy				Sell				Buy-Sell			
idx_4	μ_b	9.63×10^{-4}	σ_b	0.0113	μ_s	-8.35×10^{-4}	σ_s	0.0127	μ_{bs}	9.00×10^{-4}	σ_{bs}	0.0120
	$\widetilde{\mu}_b$	6.44×10^{-4}	$\widetilde{\sigma}_b$	0.0118	$\widetilde{\mu}_s$	-3.96×10^{-4}	$\widetilde{\sigma}_s$	0.0120	$\widetilde{\mu}_{bs}$	5.28×10^{-4}	$\widetilde{\sigma}_{bs}$	0.0119
	$\mu_b < \widetilde{\mu}_b$	0.16	$\sigma_b > \widetilde{\sigma}_b$	0.48	$\mu_s > \widetilde{\mu}_s$	0.13	$\sigma_s > \widetilde{\sigma}_s$	0.68	$\mu_{bs} < \widetilde{\mu}_{bs}$	0.10	$\sigma_{bs} > \widetilde{\sigma}_{bs}$	0.55
idx_5	μ_b	1.76×10^{-3}	σ_b	0.0143	μ_s	-1.49×10^{-3}	σ_s	0.0153	μ_{bs}	1.63×10^{-3}	σ_{bs}	0.0148
	$\widetilde{\mu}_b$	1.78×10^{-3}	$\widetilde{\sigma}_b$	0.0157	$\widetilde{\mu}_s$	-4.90×10^{-4}	$\widetilde{\sigma}_s$	0.0157	$\widetilde{\mu}_{bs}$	1.17×10^{-3}	$\widetilde{\sigma}_{bs}$	0.0158
	$\mu_b < \widetilde{\mu}_b$	0.44	$\sigma_b > \widetilde{\sigma}_b$	0.36	$\mu_s > \widetilde{\mu}_s$	0.01	$\sigma_s > \widetilde{\sigma}_s$	0.50	$\mu_{bs} < \widetilde{\mu}_{bs}$	0.11	$\sigma_{bs} > \widetilde{\sigma}_{bs}$	0.45
idx_{14}	μ_b	3.65×10^{-4}	σ_b	0.0151	μ_s	-3.36×10^{-4}	σ_s	0.0175	μ_{bs}	3.50×10^{-4}	σ_{bs}	0.0163
	$\widetilde{\mu}_b$	3.61×10^{-5}	$\widetilde{\sigma}_b$	0.0169	$\widetilde{\mu}_s$	2.76×10^{-4}	$\widetilde{\sigma}_s$	0.0170	$\widetilde{\mu}_{bs}$	-1.14×10^{-4}	$\widetilde{\sigma}_{bs}$	0.0170
	$\mu_b < \widetilde{\mu}_b$	0.26	$\sigma_b > \widetilde{\sigma}_b$	0.36	$\mu_s > \widetilde{\mu}_s$	0.13	$\sigma_s > \widetilde{\sigma}_s$	0.64	$\mu_{bs} < \widetilde{\mu}_{bs}$	0.12	$\sigma_{bs} > \widetilde{\sigma}_{bs}$	0.52
idx_{16}	μ_b	1.58×10^{-4}	σ_b	0.0141	μ_s	-3.22×10^{-4}	σ_s	0.0160	μ_{bs}	2.40×10^{-4}	σ_{bs}	0.0150
	$\widetilde{\mu}_b$	1.32×10^{-4}	$\widetilde{\sigma}_b$	0.0152	$\widetilde{\mu}_s$	-3.72×10^{-4}	$\widetilde{\sigma}_s$	0.0153	$\widetilde{\mu}_{bs}$	2.53×10^{-4}	$\widetilde{\sigma}_{bs}$	0.0153
	$\mu_b < \widetilde{\mu}_b$	0.43	$\sigma_b > \widetilde{\sigma}_b$	0.33	$\mu_s > \widetilde{\mu}_s$	0.50	$\sigma_s > \widetilde{\sigma}_s$	0.72	$\mu_{bs} < \widetilde{\mu}_{bs}$	0.51	$\sigma_{bs} > \widetilde{\sigma}_{bs}$	0.55

Table 8.19 E-GARCH aggregate bootstrap results

idx_i	Buy				Sell				Buy-Sell			
idx_4	μ_b	9.63×10^{-4}	σ_b	0.0113	μ_s	-8.35×10^{-4}	σ_s	0.0127	μ_{bs}	9.00×10^{-4}	σ_{bs}	0.0120
	$\tilde{\mu}_b$	7.16×10^{-4}	$\tilde{\sigma}_b$	0.0105	$\tilde{\mu}_s$	-3.02×10^{-4}	$\tilde{\sigma}_s$	0.0110	$\tilde{\mu}_{bs}$	5.17×10^{-4}	$\tilde{\sigma}_{bs}$	0.0107
	$\mu_b < \tilde{\mu}_b$	0.14	$\sigma_b > \tilde{\sigma}_b$	0.83	$\mu_s > \tilde{\mu}_s$	**0.03**	$\sigma_s > \tilde{\sigma}_s$	0.97	$\mu_{bs} < \tilde{\mu}_{bs}$	**0.05**	$\sigma_{bs} > \tilde{\sigma}_{bs}$	0.91
idx_5	μ_b	1.76×10^{-3}	σ_b	0.0143	μ_s	-1.49×10^{-3}	σ_s	0.0153	μ_{bs}	1.63×10^{-3}	σ_{bs}	0.0148
	$\tilde{\mu}_b$	1.57×10^{-3}	$\tilde{\sigma}_b$	0.0143	$\tilde{\mu}_s$	-6.41×10^{-4}	$\tilde{\sigma}_s$	0.0154	$\tilde{\mu}_{bs}$	1.13×10^{-3}	$\tilde{\sigma}_{bs}$	0.0149
	$\mu_b < \tilde{\mu}_b$	0.28	$\sigma_b > \tilde{\sigma}_b$	0.52	$\mu_s > \tilde{\mu}_s$	**0.06**	$\sigma_s > \tilde{\sigma}_s$	0.52	$\mu_{bs} < \tilde{\mu}_{bs}$	**0.04**	$\sigma_{bs} > \tilde{\sigma}_{bs}$	0.52
idx_{14}	μ_b	3.65×10^{-4}	σ_b	0.0151	μ_s	-3.36×10^{-4}	σ_s	0.0175	μ_{bs}	3.50×10^{-4}	σ_{bs}	0.0163
	$\tilde{\mu}_b$	-6.16×10^{-5}	$\tilde{\sigma}_b$	0.0155	$\tilde{\mu}_s$	4.64×10^{-4}	$\tilde{\sigma}_s$	0.0163	$\tilde{\mu}_{bs}$	-2.58×10^{-4}	$\tilde{\sigma}_{bs}$	0.0159
	$\mu_b < \tilde{\mu}_b$	0.16	$\sigma_b > \tilde{\sigma}_b$	0.38	$\mu_s > \tilde{\mu}_s$	**0.08**	$\sigma_s > \tilde{\sigma}_s$	0.82	$\mu_{bs} < \tilde{\mu}_{bs}$	**0.03**	$\sigma_{bs} > \tilde{\sigma}_{bs}$	0.62
idx_{16}	μ_b	1.58×10^{-4}	σ_b	0.0141	μ_s	-3.22×10^{-4}	σ_s	0.0160	μ_{bs}	2.40×10^{-4}	σ_{bs}	0.0150
	$\tilde{\mu}_b$	9.33×10^{-5}	$\tilde{\sigma}_b$	0.0141	$\tilde{\mu}_s$	-3.07×10^{-4}	$\tilde{\sigma}_s$	0.0149	$\tilde{\mu}_{bs}$	2.02×10^{-4}	$\tilde{\sigma}_{bs}$	0.0145
	$\mu_b < \tilde{\mu}_b$	0.40	$\sigma_b > \tilde{\sigma}_b$	0.51	$\mu_s > \tilde{\mu}_s$	0.45	$\sigma_s > \tilde{\sigma}_s$	0.86	$\mu_{bs} < \tilde{\mu}_{bs}$	0.43	$\sigma_{bs} > \tilde{\sigma}_{bs}$	0.75
idx_{18}	μ_b	4.56×10^{-5}	σ_b	0.0101	μ_s	1.99×10^{-4}	σ_s	0.0107	μ_{bs}	-7.11×10^{-5}	σ_{bs}	0.0104
	$\tilde{\mu}_b$	7.60×10^{-5}	$\tilde{\sigma}_b$	0.0094	$\tilde{\mu}_s$	3.09×10^{-4}	$\tilde{\sigma}_s$	0.0100	$\tilde{\mu}_{bs}$	-1.10×10^{-4}	$\tilde{\sigma}_{bs}$	0.0097
	$\mu_b < \tilde{\mu}_b$	0.55	$\sigma_b > \tilde{\sigma}_b$	0.84	$\mu_s > \tilde{\mu}_s$	0.34	$\sigma_s > \tilde{\sigma}_s$	0.79	$\mu_{bs} < \tilde{\mu}_{bs}$	0.42	$\sigma_{bs} > \tilde{\sigma}_{bs}$	0.82

Tables 8.18 and 8.19 present the aggregate results from the bootstrap methodology where a 1-day HP was used. For all examined indices returns generated by buy, sell and combined (buy-sell) signals had mean values with the expected by a technician sign: i.e. positive for buy ($\mu_b > 0$), negative for sell ($\mu_s < 0$) and positive for buy-sell ($\mu_{bs} > 0$). The only exception are positive mean daily returns for sell signals for idx_{18} which affects the combined performance of technical tools ($\mu_{bs} < 0$). In absolute terms, buy signals generate larger returns than sell signals ($|\mu_b| > |\mu_s|$) at idx_4, idx_5 and idx_{14}. This picture is reversed at idx_{16} and idx_{18}. However, considering the mean values reported in Table 8.7 the whole picture is reversed: i.e. ($|\mu_b - \overline{\mu}| < |\mu_s - \overline{\mu}|$) for indices idx_4, idx_5 and idx_{14} and ($|\mu_b - \overline{\mu}| < |\mu_s - \overline{\mu}|$) for idx_{16} and idx_{18}. Finally, on every index, following buy signals generate less volatile returns than following sell signals $\sigma_b < \sigma_s$, which is aligned with the corresponding finding reported in Brock et al. (1992).

In Table 8.18 aggregate results for the GARCH-m bootstrap methodology are reported. When TA is applied on the simulated series generated by this null model, returns generated by bearish strategies volatile more that returns generated by bullish ones at every index case, i.e. $\widetilde{\sigma}_s > \widetilde{\sigma}_b$.We treat the fractions $\sigma_b > \widetilde{\sigma}_b$, $\sigma_s > \widetilde{\sigma}_s$ and $\sigma_{bs} > \widetilde{\sigma}_{bs}$ as simulated "*p*-values" which express the fractions where buy, sell and buy-sell signals generate returns on the original price series with more volatility than those resulted by simulated series. The null model replicates well the volatility for every type of signal and in all market indices examined (all "*p*-values" fluctuate around the 0.5 level). Regarding the mean buy "*p*-values" ($\mu_b < \widetilde{\mu}_b$), they are all greater than 0.1 which indicates the significance level at 10 %. This implies that GARCH-m does generate buy returns which are consistent with the positive returns observed in the actual series. Although marginally improved, similar is the performance of bearish trading signals. The only exception is observed at idx_5 where the "*p*-value" of 0.01 constitutes bearish trading signals efficient. Overall, TA seems to have some predictability in the first three indices however results are not significant.

Aggregate results when the E-GARCH model is used as a null model are presented in Table 8.19. Again, simulated returns generated by buy signals are less volatile than those followed by sell signals, $\widetilde{\sigma}_s > \widetilde{\sigma}_b$, for all market indices. Although not significantly, E-GARCH underestimates volatilities in bullish returns at idx_4 and idx_{18} and in bearish returns at indices 14 and 16. However the corresponding fraction for idx_4 is significant at the 5 % level with a fraction $\widetilde{\sigma}_s > \widetilde{\sigma}_b$ of 0.97. This implies that for idx_4 and idx_{14} significant *p*-values reported in the bearish returns might result by the higher risk embedded in the original series compared with the simulated series. Thus significant "*p*-values" for buy-sell signals should not be considered as evidence in favour of TA. The only robust result is that reported in idx_5 which is consistent with the corresponding one reported by the GARCH-m approach.

The whole process was repeated by increasing the holding periods up to 15 days and Fig. 8.3 presents the aggregate GARCH-m simulation results. In subfigures ***a***, ***d***, ***g*** and ***j*** the evolution of $\mu_b - \widetilde{\mu}_b$, $\widetilde{\mu}_s - \mu_s$ and $\mu_{bs} - \widetilde{\mu}_{bs}$ is presented for different holding period values for idx_4, idx_5, idx_{14} and idx_{16} respectively. Sell signals seem

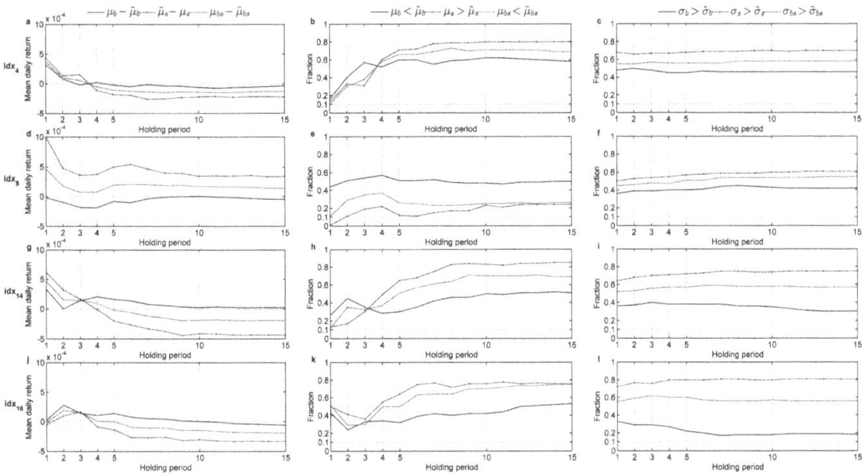

Fig. 8.3 Aggregate simulation results with GARCH-m, for different holding periods

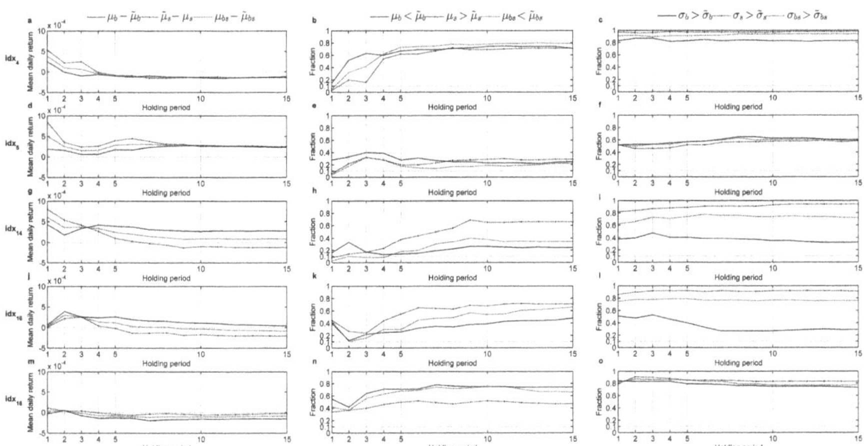

Fig. 8.4 Aggregate simulation results with E-GARCH, for different holding periods

to outperform buy signals when short holding periods are used with the exception of idx$_{16}$ where the reverse holds. This superiority is more apparent in idx$_5$ as we already pointed out previously. The second column of subfigures (***b***, ***e***, ***h*** and ***k***) reports the corresponding simulated "p-values". It is clear that the predictability of TA is negatively related with the holding period used. Finally subfigures ***c***, ***f***, ***i*** and ***l*** present the simulated "p-values" for the volatility. Obviously, increasing the number of holding periods, does not affect the volatilities' fractions. In addition, these p-values fluctuate around the 0.5 level which implies that the GARCH-m replicates efficiently buy, sell and buy-sell volatilities obtained from the original price series.

Finally, Fig. 8.4 presents the corresponding aggregate simulation results when the E-GARCH is used as a null model. Results are similar, with the exception that generally simulated series generate returns after buy, sell and buy-sell signals less volatile than those obtained from the original series. The E-GARCH model seems to reproduce the same volatility characteristics with the initial series only in the case of $idx_{5.}$

8.5 Conclusions

In this chapter the predictive performance of the technical patterns presented in Chaps. 4, 5 and 6 along with that of technical indicators presented in Chap. 7 was assessed. This set of trading rules was applied on 18 major world market indices. The efficacy of TA was tested by both implementing ordinary statistical tests and a bootstrap methodology. The first include *t*-tests and Bernoulli trials. However, while ordinary tests may assume normal, stationary, and time independent distributions, real price series distributional characteristics deviate from these assumptions presenting leptokurtic distributions, autocorrelation and heteroskedasticity. To deal with these issues, bootstrap tests are carried, whereby the performance of TA's tools observed in real series is compared with the one exhibited in simulated series. Three different null models were initially used for generating these random series; AR(1), GARCH-m and E-GARCH. However, AR(1) was not qualified to proceed to the bootstrap assessment, since it failed to produce IID increments when adopted on the original series.

Regarding ordinary statistical tests, the performance of each technical rule was assessed by examining the returns generated by bullish (buy) signals, bearish (sell) signals and their combination (Buy-Sell signals). Regarding buy signals, it was found that 71.21 % of cases generated positive returns with frequencies greater than a fair level of 50 %. Considering only significant cases at a level of 10 %, this percentage dropped to 25.38 %. Finally, considering the cases that generated mean returns significantly greater than the unconditional mean the corresponding proportion was even lower (15.15 %). Regarding sell signals, it was found that the proportion of cases that generated mean returns significantly lower than unconditional mean returns was greater than the corresponding proportion of cases that generated negative returns most times, regardless their size. This implies that even though bearish rules signal most of the times positive returns, when they signal correctly negative returns, these returns are greater in absolute values than the corresponding positive returns signaled by bullish rules. This finding can be attributed to the aforementioned black phenomenon. Overall, trading signals generated mean returns with the expected by TA sign most of times i.e., "Buy" and "Buy-Sell" signals generated positive mean returns in most cases, whereas the reverse stood for "Sell" signals. The whole analysis has been repeated with various short-term holding periods, and results showed that TA tools incorporated in this chapter have generally performed better for shorter holding periods.

It is asserted that technical patterns occur seldom but are reliable. Results of this chapter are aligned with this statement. On average technical patterns outperform indicators in a manner that they generate greater in absolute values mean trading returns. The subjectivity embedded in the identification of technical patterns, may allow for short time delays, for the incorporation of "new information" to prices. Here "new information" refers to the realization by the markets of a pattern's fully formation. Unbiased recognizers, like the proposed identification mechanisms of this book, may take benefit of such time delays. But patterns' superior performance is compensated by their low frequencies of observations. This is a very important aspect to consider since indicators produce numerous signals, which forces the mean trading returns to approximate the unconditional means.

Bootstrap results were acquired by comparing the actual trading returns and returns generated in simulated price series from two different null models; GARCH-m and E-GARCH. In this analysis only series that exhibited IID residuals were considered. Overall bootstrap results corroborate those acquired by ordinary statistical tests. However, they present even less statistical significant cases which imply that some of the distributional assumptions made in ordinary statistical tests may affect statistical inferences.

Finally, it is worth mentioning some issues that the reader should consider when interpreting reported results of this chapter. First results presented in this chapter regard only the particular dataset used, and it is not safe to generalize them. Second, parameter optimization was not performed, and parameters' values used in defining each trading rule were the most commonly used in the literature. Third, volume confirmation was not considered in generated trading signals. Although TA asserts that volume is important in confirming trading signals, it would be difficult to embed volume confirmation in bootstrapped simulated series. Finally, transaction costs were not considered in this study, which would exacerbate even further the predictive performance of TA.

References

Black F (1976) Studies of stock price volatility changes. In: Proceedings of the 1976 meetings of the Business and Economic Statistics Section, American Statistical Association, pp 177–181

Brock W, Lakonishok J, Lebaron B (1992) Simple technical trading rules and the stochastic properties of stock returns. J Finance 48(5):1731–1764

Chang PHK, Osler CL (1999) Methodical madness: technical analysis and the irrationality of exchange-rate forecasts. Econ J 109:636–661

Cramer D (1997) Basic statistics for social research. Routledge, London

Econometrics Toolbox™ User's Guide (R2011b) The MathWorks Inc.

Gehrig T, Menkhoff L (2006) Extended evidence on the use of technical analysis in foreign exchange. Int J Finance Econ 11(4):327–338

Joanes DN, Gill CA (1998) Comparing measures of sample skewness and kurtosis. J R Stat Soc Ser D 47(1):183–189

Lo AW, Mamaysky H, Wang J (2000) Foundations of technical analysis: computational algorithms, statistical inference, and empirical implementation. J Finance 55(4):1705–1765
Marshall BR, Qian S, Young M (2009) Is technical analysis profitable on US stocks with certain size, liquidity or industry characteristics? Appl Financ Econ 19:1213–1221
Nelson DB (1991) Conditional heteroskedasticity in asset returns: a new approach. Econometrica 59(2):347–370
Roll R (1977) A critique of the asset pricing theory's tests; Part 1: on past and potential testability of the theory. J Financ Econ 4(2):129–176
Ruey ST (2002) Analysis of financial time series. Financial econometrics. Wiley, New York
Sweeney R (1986) Beating the foreign exchange market. J Finance 41:163–182
Zapranis A, Tsinaslanidis PE (2012) Identifying and evaluating horizontal support and resistance levels: an empirical study on US stock markets. Appl Financ Econ 22:1571–1585. doi:10.1080/09603107.2012.663469

Chapter 9
Dynamic Time Warping for Pattern Recognition

9.1 Introduction

Technical analysis describes and analyses the formation characteristics of numerous technical patterns and their implications to the further price evolution. In Chap. 1, when describing TA we have mentioned that technical tools are designed in the spirit of identifying and classifying the repetitive trading behavior of market participants and its impact on prices. Technical patterns, which consist a category of technical tools, are specific forms of evolutions of price paths. They can be observed with different time lengths and on different price levels. Furthermore, TA tries to explain the aggregate market trading behaviour that causes these particular price patterns to occur. But, should a technician expect that patterns described in TA are the only ones? If not, how can someone identify the existence of new patterns without the need of describing the market psychology behind them? This chapter tries to answer these questions with the adoption of DTW and its modifications.

Within the context of TA, a technician's aim might be the identification of a specific query pattern within a particular price series. However, as we have already argued technical patterns can be formed with various time lengths and on different price levels. In this chapter our analysis on technical pattern recognition processes is extended, by presenting an alternative methodology, which combines two main modifications of the *Dynamic Time Warping* (DTW) algorithmic process; the *Derivative Dynamic Time Warping* (DDTW) and the *subsequence DTW*.

DTW is an algorithmic technique mainly used to find an optimal alignment between two given (time-dependent) sequences, which may vary in length, under certain restrictions (Muller 2007). First introduced in 1960s, DTW initially became popular in the context of speech recognition (Sakoe and Chiba 1978), and then in time series data mining, in particular in pattern recognition (Berndt and Clifford 1994) and similarity measurement (Fu 2011). DTW has been used in various scientific areas including finance (Wang et al. 2012; Tsinaslanidis and Kugiumtzis

P.E. Tsinaslanidis, A.D. Zapranis, *Technical Analysis for Algorithmic Pattern Recognition*, DOI 10.1007/978-3-319-23636-0_9

2014). DDTW (Keogh and Pazzani 2001) is one modification of DTW which considers only the estimated local derivatives of the data rather than then raw data themselves. This modification allows the user to apply the DTW not only to sequences that differ in length (x-axis) but also sequences that differ in price level (y-axis). However, when data used are noisy a smoothing process is prerequisite before adopting DDTW. Finally, subsequence DTW can be used to identify subsequences within a longer sequence, which are "similar" to another shorter query sequence. This chapter presents how the subsequence DDTW can be used for pattern identification by combining the aforementioned techniques.

The rest of this chapter is organized as follows: In Sect. 9.2 the basic DTW algorithm is being presented. In Sect. 9.3 the subsequence DDTW is being presented with one empirical and two simple simulation examples. Within this section it is also being described the *Savitzky-Golay* smoothing procedure and the local derivative estimation we are adopting. Finally Sect. 9.4 contains a discussion of the presented methodology and concludes.

9.2 The DTW Algorithm

DTW is an efficient scheme giving the distance (or similarity) of two sequences $Q \equiv \{q_1, q_2, \ldots, q_n, \ldots, q_N\}$ and $Y \equiv \{y_1, y_2, \ldots, y_m, \ldots, y_M\}$, where their lengths N and M may not be equal. An example of two such sequences Q and Y is given by (9.1) and (9.2):

$$q_n = \sin(x_n) + 0.2\varepsilon_n, \quad \varepsilon_n \sim IID(0,1), \quad x_n \epsilon [0, 2\pi] \quad and \quad N = 35 \tag{9.1}$$

$$y_m = \sin(x_m) + 0.2\varepsilon_m, \quad \varepsilon_m \sim IID(0,1), x_m \epsilon [0, 2\pi] \quad and \quad M = 50 \tag{9.2}$$

Clearly, both equations represent a sine with Gaussian white noise in the closed interval $[0, 2\pi]$ but with different lengths. The DTW algorithmic process begins with the calculation of the distance between any two components q_n and y_m of Q and Y, which results in forming the distance (or cost) matrix $\boldsymbol{D} \in \mathbb{R}^{N \times M}$ (Fig. 9.1b). Various distance measures can be used for this purpose, however for this simplified illustration we use the absolute value of the difference i.e., $d(q_n, y_m) = |q_n - y_m|$.

The goal is to find the optimal alignment path between Q and Y of minimum overall cost (cumulative distance). A valid path is a sequence of elements $Z \equiv \{z_1, z_2, \ldots, z_k, \ldots, z_K\}$ with $z_k = (n_k, m_k)$, $k = 1, \ldots, K$, denoting the positions in the distance matrix $\boldsymbol{D}$ that satisfy the boundary, monotonicity and step size conditions. The boundary condition ensures that the first and the last element of Z are $z_1 = (1, 1)$ and $z_K = (N, M)$, respectively (i.e. the bottom left and the top right corner of $\boldsymbol{D}$, see Fig. 9.1b). The other two conditions ensure that the path always moves up, right or up and right of the current position in $\boldsymbol{D}$, i.e. $z_{k+1} - z_k \in \{(1,0), (0,1), (1,1)\}$.

The second step of the algorithm involves the calculation of the total distance of each valid path. To do so, the cost matrix of accumulated distances $\widetilde{\boldsymbol{D}} \in \mathbb{R}^{N \times M}$ is

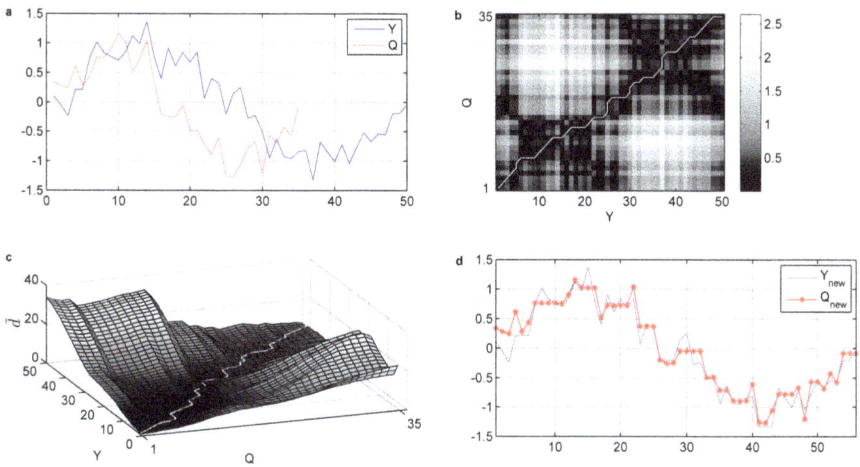

Fig. 9.1 (**a**) Q and Y price series of unequal length. (**b**) Colormap of the distance (cost) matrix. (**c**) 3D illustration of the accumulated distance (cost) matrix. (**d**) Sequences Q and Y aligned with DTW. In (**b**) and (**c**) the *white solid line* is the optimal warping path

constructed with initial condition $\widetilde{d}(1,1) = d(1,1)$. Every other element of $\widetilde{\boldsymbol{D}}$, is defined as

$$\widetilde{d}(n,m) = d(n,m) + \min\left\{\widetilde{d}(n-1,m), \widetilde{d}(n,m-1), \widetilde{d}(n-1,m-1)\right\}. \tag{9.3}$$

In the algorithm's design we set $\widetilde{d}(0,m) = \widetilde{d}(n,0) = +\infty$, for $m \in [1:M]$[1] and $n \in [1:N]$, in order to define the accumulated distances for all elements of $\widetilde{\boldsymbol{D}}$ by using (9.3) (see Fig. 9.1c). At this stage, the indexation regarding the adjacent cell with the minimum distance is kept, and then starting from $\widetilde{d}(N,M)$ the optimal path is identified backwards. In particular, if the optimal warping path is a sequence of elements $Z^* \equiv \left\{z_1^*, z_2^*, \ldots, z_k^*, \ldots, z_K^*\right\}$ with $z_K^* = (N,M)$, then conditioning on $z_k^* = (n,m)$, z_{k-1}^* is chosen as

$$z_{k-1}^* = \begin{cases} (1, m-1), if \;\; n = 1 \\ (n-1, 1), if \;\; m = 1 \\ \operatorname{argmin}\left\{\widetilde{d}(n-1,m-1), \widetilde{d}(n-1,m), \widetilde{d}(n,m-1)\right\}, & otherwise. \end{cases} \tag{9.4}$$

The process terminates when $n = m = 1$ and $z_k^* = (1,1)$ (Muller 2007). The optimal path for the example used, (9.1) and (9.2), is illustrated in Fig. 9.1b, c with the white

[1] We symbolize all integer values within the closed interval $[1, M]$ with $[1:M]$ i.e., $[1:M] \equiv [1, M] \cap \mathbb{Z}$

solid line. Having identified the optimal path, the initial sequences Q and Y are aligned by warping their time axis accordingly (Fig. 9.1d).

9.3 Subsequence Derivative DTW

As briefly described in Sect. 9.1 subsequence DTW can be used in applications where the user needs to identify subsequences within a longer sequence, Y, which are "similar" with another shorter sequence, that we call query, Q. The algorithmic process needed to follow is a modified version to the one presented in Sect.9.2. Let N and M be the lengths of the sequences Q and Y respectively, where $N \ll M$. First the distance matrix, $\boldsymbol{D} \in \mathbb{R}^{N \times M}$, is constructed as already described i.e., $d(q_n, y_m) = |q_n - y_m|$ (if the absolute difference is used as a distance measure). Then for the construction of the accumulated distance matrix, $\widetilde{\boldsymbol{D}} \in \mathbb{R}^{N \times M}$, the boundary condition is relaxed by setting $\widetilde{d}(1, m) = d(1, m)$ for all $m \in [1 : M]$ whilst the remaining elements of the matrix are filled with the process described in (9.3). In this particular case, we only need to set $\widetilde{d}(n, 0) = +\infty$ for $n \in [1 : N]$ in order to use (9.3) for calculating remaining elements of $\widetilde{\boldsymbol{D}}$.

Implementing the above modifications, actually relaxes the boundary condition while keeps the monotonicity and step size condition for deriving the optimal path. In the DTW algorithm (Sect. 9.2) valid warping paths should fulfill the boundary condition i.e., $z_1 = (1, 1)$ and $z_K = (N, M)$. In the subsequence DTW algorithm the boundary condition becomes; $z_1 = (1, a)$ and $z_K = (N, b)$, where $a \in [1 : M]$, $b \in [1 : M]$ and $1 \leq a \leq b \leq M$. By the manner $\widetilde{\boldsymbol{D}}$ was constructed, $\widetilde{d}(N, m)$ includes the total costs of all M optimal warping paths, where each one ends at time $b = m$, $m \in [1 : M]$. In other words, $\widetilde{d}(N, m)$ can be viewed as a cost function, and the user can set a cost threshold $\tau \in \mathbb{R}$ in order to identify the set $B^* = \{b_1^*, b_2^*, \ldots, b_\ell^*, \ldots, b_L^*\}$ with the L most similar subsequences' ends (9.5).

$$B^* = \arg\left(\widetilde{d}(N, b) < \tau\right), b \in [1 : M] \tag{9.5}$$

The backwards process described in (9.4) can be used to find the optimal warping paths and the starting points, a_ℓ^*, for each b_ℓ^*. However, this process may result in finding multiple paths that differ marginally only in the b_ℓ^*. For this reason, before adopting (9.5), we could use the $RW(\cdot)$ presented in Chap. 2 on the $\widetilde{d}(N, m)$ to find local minima excluding neighborhoods of b_ℓ^* from further consideration.

Consider the following example. We construct the longer sequence $Y \equiv \{y_1, y_2, \ldots, y_m, \ldots, y_M\}$, $M = 1000$, with (9.6) and (9.7):

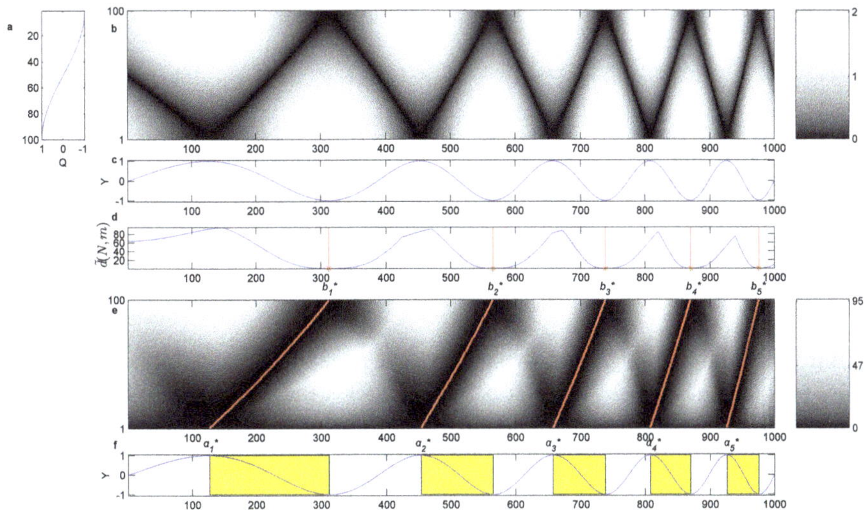

Fig. 9.2 (**a**) The query, Q, series, (**b**) the cost matrix $\boldsymbol{D}$, (**c**) the longer, Y, series, (**d**) identification of five b_ℓ^* on $\widetilde{d}(N,m)$ after adopting a threshold $\tau = 2$ and a rolling window of 20 observations, (**e**) the accumulated cost matrix $\widetilde{\boldsymbol{D}}$ with the five identified optimal warping paths, (**f**) identification of similar subsequences of Y

$$y_m = \sin(x_m) \tag{9.6}$$

$$x_m = e^{t_m}\pi,\ t_m \in [\ln(2), \ln(12)] \tag{9.7}$$

In (9.7) t_m is linearly spaced within the closed interval $[\ln(2), \ln(12)]$ and this implies that $x_m \in [2\pi, 12\pi]$. However x_m is an increasing natural exponential function which increases faster as t_m increases. Thus, considering also that a sine function has a period of 2π and that $x_m \in [2\pi, 12\pi]$, (9.6) generates a "decreasing-period" sine of 5 periods. Subsequently, we define the query sequence, $Q \equiv \{q_1, q_2, \ldots, q_n, \ldots, q_N\}$, as $q_n = \sin(x_n)$ where x_n consists of 100 values, $N = 100$, linearly spaced within the closed interval $[0.5\pi, 1.5\pi]$. Figure 9.2 illustrates the above process. First, the cost matrix (Fig. 9.2b) is constructed by calculating the absolute difference between each pair of elements of the series Q (Fig. 9.2a) and Y (Fig. 9.2c). Then, the accumulated cost matrix (Fig. 9.2e) is constructed by adopting the necessary adjustments for the subsequence DTW version. Figure 9.2d presents the elements of the accumulated cost matrix located at the upper row of $\widetilde{\boldsymbol{D}}$, i.e., the cost function $\widetilde{d}(N,b)$. In addition, five ending points $\{b_1^*, b_2^*, b_3^*, b_4^*, b_5^*\}$ of the five most similar to Q subsequences of Y are highlighted by adopting (9.5), with a threshold $\tau = 2$ and a rolling window of 20 observations in the $RW(\cdot)$. Subsequently, in Fig. 9.2e, five optimal paths with their starting points $\{a_1^*, a_2^*, a_3^*, a_4^*, a_5^*\}$ are identified by (9.4). Finally, Fig. 9.2f highlights the five similar subsequences identified in Y.

Table 9.1 Results of Fig. 9.2

ℓ	$a_\ell^*\left(x_{a_\ell^*}\right)$	$b_\ell^*\left(x_{b_\ell^*}\right)$	$b_\ell^* - a_\ell^* + 1\left(x_{b_\ell^*} - x_{a_\ell^*}\right)$	$\widetilde{d}\left(N, b_\ell^*\right)$
1	127 (2.51π)	312 (3.49π)	186 (0.9865π)	0.9395
2	454 (4.51π)	565 (5.50π)	112 (0.9928π)	0.5689
3	658 (6.50π)	738 (7.50π)	81 (1.0026π)	0.6248
4	808 (8.50π)	870 (9.50π)	63 (1.0002π)	0.7960
5	926 (10.51π)	976 (11.49π)	51 (0.9859π)	0.9799

Information regarding these five subsequences is tabulated in Table 9.1. Recall that the query sequence consists of 100 values generated by a sine function, $q_n = \sin(x_n)$, where x_n is linearly spaced within the closed interval $[0.5\pi, 1.5\pi]$. In addition, the longer sequence is given by $y_m = \sin(x_m)$, where x_m is not linearly spaced within $[2\pi, 12\pi]$ (9.7). Given this, we expect that the subsequence DTW algorithm would identify five subsequences of Y that correspond to $\left\{\left(x_{a_\ell^*}, x_{b_\ell^*}\right)\right\} = \{((2\ell + 0.5)\pi, (2\ell + 1.5)\pi)\}, \ell \in [1:5]$. Apparently, in this simulation example, the subsequence DTW algorithm identifies all these five subsequences. Finally it is worth to mention that for an identified similar subsequence, $y_{[a_\ell^*:b_\ell^*]}$, its warping cost, $\widetilde{d}\left(N, b_\ell^*\right)$, takes lower values when its length, $\left(b_\ell^* - a_\ell^* + 1\right)$, is closer to the length of the query sequence, N.

We repeated the previous experiment with the difference that this time noise is added to the series Q and Y. In particular, $q_n = \sin(x_n)$ becomes

$$q_n = \sin(x_n) + 0.2\varepsilon_n, \quad \varepsilon_n \sim IID(0, 1)\,, \tag{9.8}$$

where again x_n, $n \in [1:N]$, consists of 100 values ($N = 100$), linearly spaced within the closed interval $[0.5\pi, 1.5\pi]$ and (9.6) becomes

$$y_m = \sin(x_m) + 0.2\varepsilon_n, \quad \varepsilon_n \sim IID(0, 1), \tag{9.9}$$

keeping the definition of x_m same as in (9.7). Before proceeding to the subsequence DTW implementation series Q and Y are scaled to $[-1, 1]$. Figure 9.3 and Table 9.2 present the identification of five similar subsequences. Apparently, introducing noise to the series under consideration increases the warping cost of the identified subsequences. For this reason, a greater value should be set for the cost threshold τ (in this example $\tau = 20$). Comparing Tables 9.1 and 9.2 shows that noise affects also the algorithms accuracy since identified subsequences differ more from the desired $\left\{\left(x_{a_\ell^*}, x_{b_\ell^*}\right)\right\} = \{((2\ell + 0.5)\pi, (2\ell + 1.5)\pi)\}$, $\ell \in [1:5]$. Finally, noise seems to affect more heavily the warping cost than the length of the identified subsequence. For instance, subsequences $y_{[a_\ell^*:b_\ell^*]}$ for $\ell = 3, 4$ and 5 realise lower

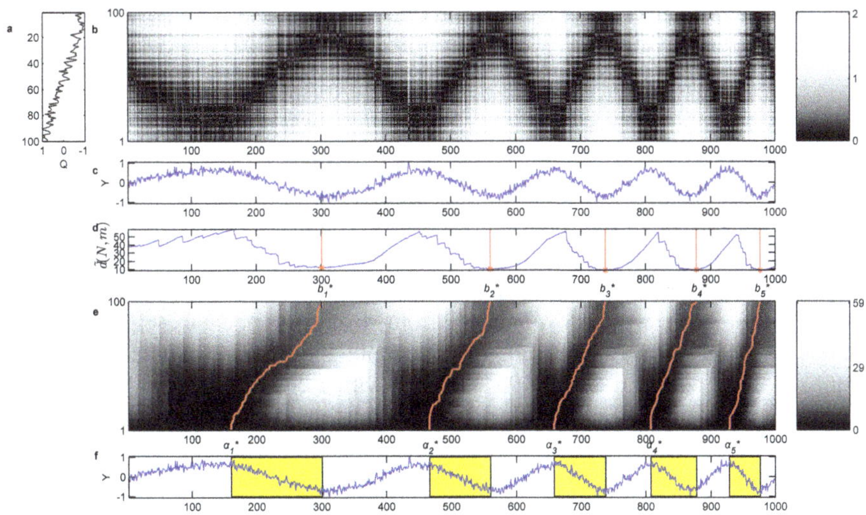

Fig. 9.3 As for Fig. 9.2 but for noisy series and adopting a threshold $\tau = 20$

Table 9.2 Results of Fig. 9.3

ℓ	$a_\ell^*\left(X_{a_\ell^*}\right)$	$b_\ell^*\left(X_{b_\ell^*}\right)$	$b_\ell^* - a_\ell^* + 1\left(X_{b_\ell^*} - X_{a_\ell^*}\right)$	$\widetilde{d}\left(N, b_\ell^*\right)$
1	161 (2.66π)	301 (3.43π)	141 (0.7606π)	12.4311
2	467 (4.61π)	560 (5.45π)	94 (0.8374π)	10.2307
3	659 (6.51π)	738 (7.50π)	80 (0.9909π)	8.9754
4	808 (8.50π)	878 (9.64π)	71 (1.1376π)	9.1690
5	928 (10.55π)	977 (11.52π)	50 (0.9688π)	8.8536

warping cost than the second similar subsequence, $y_{\left[a_2^*:b_2^*\right]}$, although the later subsequence's length is closer to the queries length.

As already described, the DTW is an algorithmic technique mainly used in finding the optimal alignment between two time series that differ in the x-axis but evolve within similar y-axis levels. However, financial price series evolve in different price levels and thus attempts of implementing the DTW algorithm (or the subsequence DTW) directly on prices might be problematic. Thus, in view of financial applications we are using the derivative DTW (DDTW) to deal with this problem (Keogh and Pazzani 2001). In this algorithmic version, the first derivatives of series Q and Y are calculated first before implementing the DTW.

However, when series are noisy, smoothing is a prerequisite step before estimating the derivatives. For instance, Keogh and Pazzani (2001) used exponential smoothing (Mills 1990). However, various smoothing techniques can be used alternatively. Orthogonal series expansion, projection pursuit, nearest-neighbor

Table 9.3 Coefficients of a cubic, local polynomial fitting with 21 data points used for the smoothing

i	0	1	2	3	4	5	6	7	8	9	10
$3{,}059 \times C_i$	329	324	309	284	249	204	149	84	9	−76	−171

estimators, average derivative estimators, splines, and artificial neural networks are examples of smoothing techniques mentioned in Campbell et al. (1997).

Subsequently an example of implementing subsequence DDTW on a real price series is being presented. Closing prices of S&P 500 for the period 12-Jun-2009 until 28-Dec-2012 were downloaded from Bloomberg database, forming a sequence of 894 observations. In this example the *Savitzky-Golay* smoothing procedure (Savitzky and Golay 1964) is used with a cubic polynomial model and a window size of $w_s = 21$ days (approximately one trading month). Let $P \equiv \{p_1, p_2, \ldots, p_m, \ldots, p_M\}$ be the aforementioned sequence of , $M = 894$ observations. Smoothed values, $\hat{p}_m$, generated from the local polynomial fit are given by (9.10).

$$\hat{p}_m = \sum\nolimits_{i=-(w_s-1)/2}^{(w_s-1)/2} C_i p_{m+i}, \quad \forall m \in \left[\frac{w_s+1}{2} : M - \frac{w_s+1}{2}\right] \tag{9.10}$$

Here, w_s is the window size (odd number) used for implementing the smoothing, and C_i, $i \in \left[-\frac{(w_s-1)}{2} : \frac{(w_s-1)}{2}\right]$ are the coefficients of the polynomial fit which are symmetric i.e., $C_i = C_{-i}$. Thus Table 9.3 reports only positive coefficients for the particular parameters used; $w_s = 21$ and a cubic polynomial (for example, $C_0 = 329/3059$, $C_1 = C_{-1} = 324/3059$ etc.). Regarding treatment of first and last $(w_s - 1)/2$ data points (in our case first and last 10 points) their smoothed values are those derived from fitting a cubic polynomial on the first and last 21 points respectively.

Subsequently, the first derivative $\{\hat{p}'_m\}$ of the smoothed sequence $\{\hat{p}_m\}$ is calculated by the estimate used in Keogh and Pazzani (2001).

$$\hat{p}'_m = \frac{(\hat{p}_m - \hat{p}_{m-1}) + (\hat{p}_{m+1} - \hat{p}_{m-1})/2}{2} \tag{9.11}$$

Keogh and Pazzani (2001) state that empirically this estimate is more robust to outliers than any estimate which considers only two datapoints. Finally, $\{\hat{p}'_m\}$ is standardized to have a mean of zero and a standard deviation of one. The resulted sequence is the longer sequence Y to be used as the first input in the sequence DTW algorithm. Let Q be the second input for the algorithm, which is the subsequence of Y composed by the last two trading months of the initial dataset (period from 1-Oct-2012 until 28-Dec-2012, $N = 61$ observations). This implies that our goal

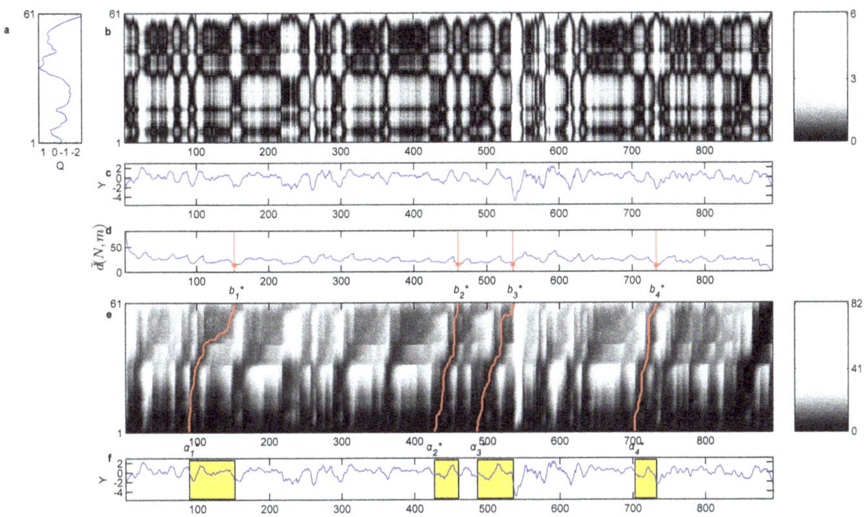

Fig. 9.4 As for Fig. 9.2 but for S&P 500. A cost threshold $\tau = 15$ and a rolling window of 20 observations is used

is to identify subsequences of Y which are similar to the query subsequence Q. With a threshold cost value $\tau = 15$ and a rolling window of 20 observations to exclude neighborhoods of b_ℓ^* from further consideration the subsequence DTW algorithm identifies four similar subsequences. Figure 9.4 presents the whole process.

Note that in Fig. 9.4d $\tilde{d}(N, M) = 0$ as expected, since in this example the Q sequence is actually the last part of Y, and although not highlighted, the algorithm identifies this part being identical to Q with a zero optimal warping cost. A closer look in Fig. 9.4f shows that there might be a pathological alignment between the first identified similar subsequence, $y_{[a_1^*:b_1^*]}$, and the query sequence. By the term pathological, we mean that this particular optimal warping path deviates strongly from the diagonal drawn from its starting and ending point (elements $\tilde{d}(1, a_1^*)$ and $\tilde{d}(N, b_1^*)$ of the accumulated cost matrix). In particular, the warping path goes initially (almost) vertically before reaching its ending point. This implies that in the alignment process, almost the first half observations of Q are matched with only the first few observations of $y_{[a_1^*:b_1^*]}$. To prevent such alignments the user should consider using global constraints within the subsequence DDTW process, like the *Sakoe-Chiba band* (Sakoe and Chiba 1978) or the *Itakura parallelogram* (Itakura 1975).

The corresponding four identified subsequences on the original S&P 500 series are illustrated in Fig. 9.5. It is clear that by using the derivative version of the subsequence DTW the algorithm identifies similar historical subsequences that differ both in time length and in price level.

Fig. 9.5 Identified similar subsequences of the S&P 500 series with the subsequence DDTW algorithm. Similar subsequences *highlighted* correspond to those identified in Fig. 9.4

9.4 Conclusions

With the methodology presented in this chapter, the reader should be able to trace similar sequences (or subsequences) to a query series of his own preference. Although the subsequence DDTW described in this chapter identifies successfully patterns regardless their time length or price level we would like to briefly discuss on some worth noted issues, and propose some interested, to our point of view, future areas for research.

First, in pattern recognition procedures, smoothing is a common prerequisite process. Each smoothing process has its own user defined parameters which are crucial in the smoothing and subsequently in the pattern recognition process. For instance, in the methodology presented in this chapter, before estimating the local derivative and adopting the subsequence DDTW we used as a smoothing device a local, cubic polynomial fitting that considers 21 data points on a rolling basis. Obviously by using other degrees for the polynomials or different rolling windows' widths would affect the whole process. Another example is the kernel regression used in Lo et al. (2000) where the bandwidth was the user defined parameter that affected the smoothing level. Lo et al. initially used the method of cross validation to find the optimal bandwidth that minimized the mean squared error of the kernel estimator. However, they found that the optimal bandwidth suggested by this approach smoothed extensively the original series. They had to rely on polling professional technical analysts to reach an acceptable solution and use only the 30 % of the optimal bandwidth resulted by the cross validation method. So the smoothing methodology that an analyst would follow along with its parameters is crucial in the pattern recognition process and should be carefully selected. We also believe that this is a really interesting area for future research.

Second, one of the main drawbacks of the DTW approach is its computational expensiveness. Although, research has been focused on this problem and different methods have been proposed to speed up the DTW matching process [see Fu (2011) and the references therein], the complexity is still an issue, especially when the DTW (or any of its modifications) is to be combined with another also computationally expensive approach. Indicatively, consider the example whereby a prediction mechanism based on the DTW algorithm is developed, and we want to assess its predictive performance with a bootstrap approach.

Third, prediction schemes could be designed based on DTW. For instance, recall that in this chapter, in the S&P 500 example we used the subsequence DDTW algorithm to find historical subsequences similar to the last part of a price series. Could we predict the future price evolution by looking at the price paths following the identified similar subsequences? That would also be a very interesting area for future research.

Overall, in view of technical pattern recognition applications DTW modifications can be used in various ways. For instance, a technician might compare and find stock price series that behave similar to a query series. In addition, by setting the query series to be a celebrated technical pattern, like the head-and-shoulders pattern, the user can aim at identifying this pattern within longer series by using subsequence DTW modifications. Finally, the DTW can be used as a similarity measure for comparing pair-wisely subseries within a longer series. The benefit, of using DTW as a similarity measure can be traced in cases where the candidate series differ in length whereby the implementation of traditional correlation measures (like the Pearson's r, the Spearman's ρ, Kendall's τ and Kruskal's Γ correlation coefficients) in not possible. Implications of this characteristic in finance applications are significant, since DTW can be used to study market seasonalities by comparing the dynamics of returns series evolutions across different months which might differ in length.

References

Berndt DJ, Clifford J (1994) Using dynamic time warping to find patterns in time series. In: Association for the Advancement of Artificial Intelligence, Workshop on Knowledge Discovery in Databases (AAAI), pp 229–248

Campbell JY, Lo AW, MacKinlay AC (1997) The econometrics of financial markets. Princeton University Press, New Jersey

Fu TC (2011) A review on time series data mining. Eng Appl Artif Intell 24(1):164–181

Itakura F (1975) Minimum prediction residual principle applied to speech recognition. IEEE Trans Acoust Speech Sig Process ASSP-23:52–72

Keogh EJ, Pazzani MJ (2001) Derivative dynamic time warping. Paper presented at the SIAM International Conference on Data Mining

Lo AW, Mamaysky H, Wang J (2000) Foundations of technical analysis: computational algorithms, statistical inference, and empirical implementation. J Financ 55(4):1705–1765

Mills TC (1990) Time series techniques for economists. Cambridge University Press, Cambridge, UK

Muller M (2007) Information retrieval for music and motion. Springer, Heidelberg
Sakoe H, Chiba S (1978) Dynamic programming algorithm optimization for spoken word recognition. IEEE Trans Acoust Speech Sig Process ASSP-26(1):43–49
Savitzky A, Golay MJE (1964) Smoothing and differentiation of data by simplified least squares procedures. Anal Chem 36(8):1627–1639
Tsinaslanidis PE, Kugiumtzis D (2014) A prediction scheme using perceptually important points and dynamic time warping. Expert Syst Appl 41(15):6848–6860
Wang GJ, Xie C, Han F, Sun B (2012) Similarity measure and topology evolution of foreign exchange markets using dynamic time warping method: Evidence from minimal spanning tree. Physica A 391:4136–4146